Internetworking Over ATM: An Introduction

Brian Dorling
Jaap Burger
Daniel Freedman
Chris Metz

INTERNATIONAL TECHNICAL SUPPORT ORGANIZATION
RESEARCH TRIANGLE PARK, NORTH CAROLINA 27709

PRENTICE HALL PTR
UPPER SADDLE RIVER, NEW JERSEY 07458

TAKE NOTE!
Before using this information and the product it supports, be sure to read the general information in Appendix H, "Special Notices" on page 227.

First Edition (November 1996)

This edition applies to Asynchronous Transfer Mode and related protocols.

Comments may be addressed to:
IBM Corporation, International Technical Support Organization
Dept. HZ8 Building 678
P.O. Box 12195
Research Triangle Park, NC 27709-2195

Published by Prentice Hall PTR
Prentice-Hall, Inc.
A Simon & Schuster Company
Upper Saddle River, NJ 07458

Editorial/Production Supervision: Joe Czerwinski

Acquisitions Editor: Michael E. Meehan

Editorial Assistant: Kate Hargett

Manufacturing Manager: Alexis R. Heydt

Cover Design: Design Source

Cover Design Director: Jerry Votta

The publisher offers discounts on this book when ordered in bulk quantities.
For more information, contact:

Corporate Sales Department
Prentice Hall PTR
One Lake Street
Upper Saddle River, NJ 07458
Phone: 800-382-3419
FAX: 201-236-7141
E-mail: corpsales@prenhall.com

Printed in the United States of America

10 9 8 7 6 5 4 3 2

ISBN 0-13-612384-8

Prentice-Hall International (UK) Limited, *London*

Prentice-Hall of Australia Pty. Limited, *Sydney*

Prentice-Hall Canada Inc., *Toronto*

Prentice-Hall Hispanoamericana, S.A., *Mexico*

Prentice-Hall of India Private Limited, *New Delhi*

Prentice-Hall of Japan, Inc., *Tokyo*

Simon & Schuster Asia Pte. Ltd., *Singapore*

Editora Prentice-Hall do Brasil, Ltda., *Rio de Janeiro*

Contents

Preface

For the foreseeable future a significant percentage of devices using an ATM network will do so indirectly, and will continue to be directly attached to ″legacy″ media (such as Ethernet and token ring). In addition these devices will continue to utilize ″legacy″ internetwork layer protocols (for example, IP, IPX, APPN, etc). This means that in order to effectively use ATM, there must be efficient methods available for operating multiple internetwork layer protocols over heterogeneous networks built from ATM switches, routers, and other switched devices. This challenge is commonly referred to as the operation of multiprotocol over ATM.

This book reviews the various options for the transport and support of multiple protocols over ATM.

This book was written for networking consultants, systems specialists, system planners, network designers and network administrators who need to learn about SVN and associated protocols in order to design and deploy networks that utilize components from this framework. It provides readers with the ability to differentiate between the different offerings. A working knowledge of ATM is assumed.

It is intended to be used with *High-Speed Networking Technology: An Introductory Survey*, which describes methods for data transmission in high speed networks, and *Asynchronous Transport Mode (ATM) Technical Overview*, which describes ATM, a link-level protocol using the methods described in *High-Speed Networking Technology: An Introductory Survey* to transmit various types of data together over the same physical links. This book describes the networking protocols that use ATM as the underlying link level protocol.

How This Book Is Organized

- Chapter 1, "Introduction"

  This chapter introduces the reader to high speed networking. It reviews current trends in networking technology, discusses switching as an alternative to routing, why existing networking protocols don′t work well at higher speeds, and gives on short description of ATM.

- Chapter 2, "Emulated and Virtual LANs"

  A brief introduction to ATM Forum Compliant LAN Emulation which is one of the underlying technologies used by SVN is given here.

- Chapter 3, "Multiprotocol over ATM (MPOA)"

This chapter gives an description of the multiprotocol over ATM (MPOA) standard as it stands at this moment in time. MPOA is designed to allow other protocols to be transported over ATM connections.

- Chapter 4, "APPN Support in ATM Networks"

  IBM's Advanced peer to peer networking is being enhanced to give it native access to ATM connections. An overview of the architecture that changes that the APPN Implementers Workshop (AIW) plans to adopt is given in this chapter.

- Chapter 5, "IP Support in ATM Networks"

  This chapter explains various solutions for running IP over ATM networks. IP address resolution, next-hop resolution and IP multicasting are discussed.

- Chapter 6, "PNNI Phase 1 and Integrated PNNI"

  Various solutions to running legacy protocols over IP networks rely on overlay models, overlaying an existing protocol onto an ATM network, without truly integrating the two. In this chapter the protocol for routing of a call setup in an ATM network, private network-to-network interface (PNNI) is explained, then PNNI augmented routing (PAR) and integrated PNNI (I-PNNI) are discussed as methods to fully integrate IP routing into ATM networks.

- Chapter 7, "Switched Virtual Networking Architecture"

  This chapter gives a brief overview of IBM's Switched Virtual Networking (SVN) Architecture. The functional components that make up an SVN network and the underlying technologies are described here.

- Chapter 8, "The Future of IP"

  The next-generation IP protocol (IPv6), the Integrated Services model, and Resource Reservation Protocol are covered in this chapter.

- Appendix A, "Protocol Stack Reference"

  This appendix contains diagrams of the ATM, SNA and IP protocol stacks.

- Appendix B, "ATM Service Categories"

  This appendix describes ATM service categories.

- Appendix C, "AAL Service Classes"

  This appendix describes AAL service classes.

- Appendix E, "Multiprotocol Encapsulation over AAL 5 (RFC 1483)"

  This appendix contains a short description of methods of multiprotocol encapsulation over AAL type 5.

- Appendix F, "Cells in Frames (CIF)"

The cells in frames technology, a method of transporting ATM cells over ethernet, is described here.

- Appendix G, "Server Cache Synchronization Protocol (SCSP) -NBMA"

  The server cache synchronization protocol, a protocol that will probably be used by various other technologies described in this book, is introduced here.

- Appendix D, "ATM Address Formats"

  This appendix contains a reference of ATM addressing formats.

The Team That Wrote This Book

This book was produced by a team of specialists from around the world working at the Systems Management and Networking ITSO Center, Raleigh.

Brian Dorling is IBM's technical liason and support for communications architectures at the ITSO Raleigh Center. Brian is responsible for a broad range of IBM communication architectures including Advanced Peer-to-Peer Networking (APPN), Multiprotocol Transport Networking (MPTN), Networking Broadband Services (NBBS), and Switched Virtual Networking (SVN). After joining IBM in 1978, Brian has worked as a Customer Engineer and Systems Engineer in the networking field in the UK and Germany.

Daniel Freedman is a senior technical specialist providing both pre- and post- sales technical (and marketing) support. He has five years of experienc in networking, encompassing the design, implementation and tuning of (complex) LAN and multiprotocol based router networks. During the past two years he has specialized in the area of campus ATM networking.

Chris Metz is a Certified I/T Specialist for the IBM Corp. He specializes in ATM and multiprotocol routing, and possesses a detailed knowledge of network architectures and technologies gained in over 12 years of service as an IBM technical specialist and consultant. He is the author of several technical papers detailing ATM network protocols, as well as "ATM and Multiprotocol Networking", published by McGraw-Hill this year. He holds a B.A. in mathematics from Rollins College and is a member of ACM/Sigcomm and IEEE.

Jaap Burger joined IBM in 1986 and since then has been involved in almost every field of networking, most recently network design and ATM. He is the author of redbooks about frame relay implementations and performance guidelines.

Thanks to the following people for their invaluable contributions to this project:

Gary Dudley
IBM Research Triangle Park

Harry Dutton
IBM Australia

John G. Waclawsky Ph.D
IBM Gaithersburg

Peter C. Wong
IBM Bethesda

The Editing and Graphics Team
ITSO Raleigh

Chapter 1. Introduction

Before we discuss the protocols involved in internetworking over ATM networks in detail, it is useful to review some of the challenges involved in high-speed networking.

1.1 Networking Evolution

Today's networking environment is characterized by a diverse mix of topologies, geographic spans, carrier services, equipment, interfaces, physical media, and transmission speeds. While many business processes and applications were and still are well-served by the networking technologies that support this environment, new applications and expanded communications requirements for existing applications are creating a whole new set of networking challenges.

Users are becoming increasingly sophisticated and demanding in their expectations of computing services, thereby stimulating the development of increasingly complex applications that draw from diverse forms of information, for example, voice, video, and data, and hence are termed multimedia. These emerging multimedia applications will not only serve to entertain their users, but will also provide the competitive edge that business constantly seeks. Taken together, these observations suggest that the networking demands of the near future cannot be met by a simple extrapolation of the current networking methods and products.

The requirements of the emerging networking environment present a multifaceted challenge as follows:

- To provide dynamic and transparent connectivity to join wide area and local area domains as well as to integrate private and public subnetworks. Any given enterprise may find it desirable or even necessary to create a network infrastructure consisting of some combination of local and wide area, and public and private components. The key to the success of this hybrid network is the ability to manage it across all segments.
- To support applications with unprecedented needs for high-transmission speeds, large transmission capacities, the flexibility of bandwidth on demand, end-to-end quality of service, and effective network management capabilities.
- To offer products that enable the creation of networking solutions characterized by lower operational costs and better price-to-performance ratios.

Because of the inherent complexity resulting from mixed technologies, it becomes extremely difficult to satisfy all these requirements with existing networking tools. For example, in order for an Ethernet-attached client in a branch office to access an

FDDI-attached server in a different location, at least three different networking protocols must be made to communicate. While it is clear that this scenario will exist for some time to come, the goal is to provide a technology that will not only provide the desired simplification, but will also support those emerging applications that will demand end-to-end quality of service and bandwidth reservation capability. Asynchronous transfer mode (ATM) provides the foundation for this technology.

1.2 The Challenges of High-Speed Networking

Designing an architecture for high-speed networks presents two kinds of challenges. The first is related to the nature of the technology itself; the second kind is related to the problems of integrating voice, video, image, and data transfers in the same network.

With respect to the changes in transmission technology (that we transmit data at ever faster speeds), two basic facts become important:

1. Processing time available at intermediate nodes decreases.
2. The propagation delay remains constant and, consequently, the product of bandwidth and propagation delay increases, which means that the amount of data *in flight* increases.

In addition, as the installed physical transmission media are changed from copper to fiber, the effective error rates of the links decrease dramatically.

1.2.1.1 Processing at Intermediate Network Nodes

> **Note**
>
> Perhaps the fundamental challenge for high-speed networking is to minimize the processing time for each packet or cell within intermediate nodes in the network.

For example, the requirement for a telephone user is to ensure that the end-to-end propagation delay should not be larger than 100 ms. This will be apparent to anyone who has experienced a long distance phone conversation via satellite with a propagation delay significantly more than 100 ms. This means you have a budget of 100 ms to deliver a voice packet between two end users. There is a 16 ms packetization delay[1] at the sending side and a 20 ms delay at the receiving side. About 20 ms is the unalterable propagation delay to get the data across, for example, the Atlantic. That means, there remains 44 ms for all intranode processing in all intermediate nodes as the packet is

1 The calculations are based on 128-byte packets for voice data; for smaller packet sizes the packetization and playout delays will be shorter.

navigated through the network. On a 5-hop transmission path, each node would have about 8 ms for processing all the protocols necessary to forward the packet to the next link, including any queuing time in front of the link. Given that networks may contain 10 hops, the network design point should probably be below 4 ms.

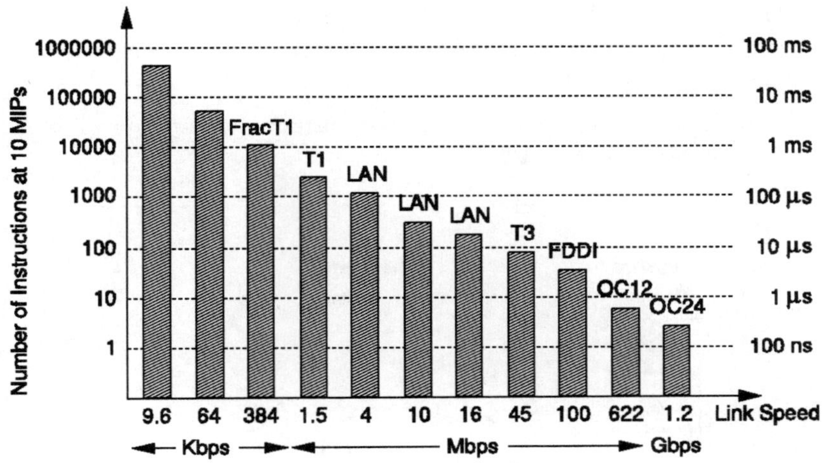

Figure 1. Available Node Processing Time for a 53-Byte ATM Cell

Note: The scales on both axes of the graph are logarithmic.

Figure 1 shows the amount of time that a processor has per 53-byte ATM cell received on links of different speeds. On the left of the figure, a scale shows the number of instructions that a 10-MIPS processor would have per cell to make a switching decision before the next cell arrives. Typical packet switches take an average of between 1000 and 3000 instructions to receive a packet, transfer it to an output link, and then start transmitting it towards its destination.

Looking at it from the receiving node's perspective (where links are full-duplex and a node is sender and receiver on a link at the same time), the graph shows that our 10-MIPS processor can execute 2826 instructions for each cell received on a T1 link (to decide where to send the cell next and to handle all protocols) before the next cell arrives on that link. On an OC-24 link, however, our fictional 10-MIPS machine could execute no more than 3 instructions per cell. Of course, nobody will use a 10-MIPS processor to feed a 1.2-Gbps link, but even with 50 MIPS you only have 17 instructions to keep your link busy. In addition, an intermediate switch node will have not only one or two links, but 10, 20 or 100.

The implications of this demand to reduce intermediate node processing time lead us to the following conclusions, as depicted in Figure 2 on page 4:

- The architecture (any architecture) must minimize the awareness and the function of intermediate nodes.
- There can be no hop-by-hop flow control or error recovery (at least for applications demanding high bandwidth).
- Congestion and flow control have to be provided at the endpoints of network connections.
- The architecture should allow intermediate node functions, wherever possible and feasible, to be implemented in hardware.

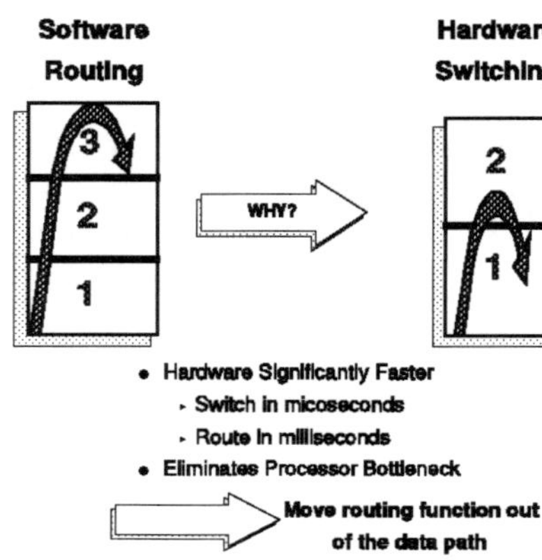

Figure 2. *Implications to Reduce Intermediate Node Processing Time*

To meet the above requirements networks will need to have a hardware-controlled switching function.

There is no way that current software-based switched architectures can deliver a fraction of the required throughput, even assuming much faster processors.

From a different viewpoint, we can see that if we continue to use a software-based switched architecture, although we may be able to integrate a high-speed interface into the device, it will not be possible to drive the interface to capacity.

1.2.1.2 In-Flight Data

Congestion control becomes more complicated as the amount of data *in flight* (that is, the product of propagation delay and nominal transmission speed of a link) increases. For example, if the propagation delay is 20 ms, the number of bytes in flight on a T1 link (1.5 Mbps) is about 3000, which increases to 3000000 bytes on an OC-24 link. The main consequence of having a large amount of in-flight data is that reactive flow control mechanisms are no longer effective in high-speed networks; by the time sources receive a congestion indication, it may be too late to react. High-speed networks (especially wide area networks with the associated huge amount of in-flight data) need a rate-based, preventive congestion control algorithm that controls the rate at which a source may send data into the network.

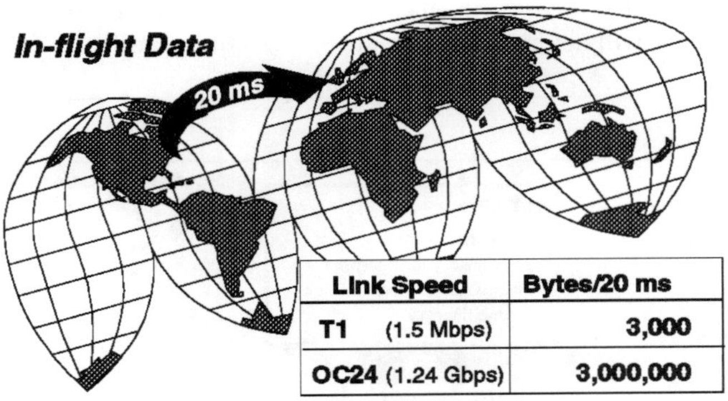

Link Speed	Bytes/20 ms
T1 (1.5 Mbps)	3,000
OC24 (1.24 Gbps)	3,000,000

Figure 3. Data In Flight

1.2.2 The Changing Role of Routing

Early LANs had a very limited geographical scope. Later, bridges, and then routers, were used to connect multiple physical LANs in an establishment so that a device on one LAN could communicate with a device on a different LAN. Bridges simply pass on indiscriminately (that is, broadcast) messages that are not addressed to a station on the originating LAN to the next LAN. Routers, on the other hand, interpret the address of those messages not designated for a station on the local LAN and forward each message to the router that owns the connection to the destination. This approach reduces the number and size of broadcasts, thus avoiding broadcast storms that can totally degrade the performance of bridge-based networks. When it became necessary to link LANs at multiple sites, a connection to a wide area network was needed, and it was most often accomplished by routers.

The role of the router in a data network has been essential to the establishment of integrated subnetworks as well. For administrative purposes, a given protocol like TCP/IP may be subnetted, producing separate logical networks. A router must be employed for such subnets to retain their identity across multiple physical LANs.

Routing normally works well for nondelay sensitive data traffic, but routing SNA traffic over IP networks where there is no prioritization gives unacceptable performance since service level guarantees are not possible. Moreover, since voice and video are delay-sensitive, their addition to the network will stress the router model beyond its capabilities. In addition, the increase in network traffic volume is creating the need for higher bandwidth, which the packet-forwarding model of router processing can rarely provide effectively. For these reasons, the traditional large centralized routers, as depicted in Figure 4, will not be able to handle the resulting workload.

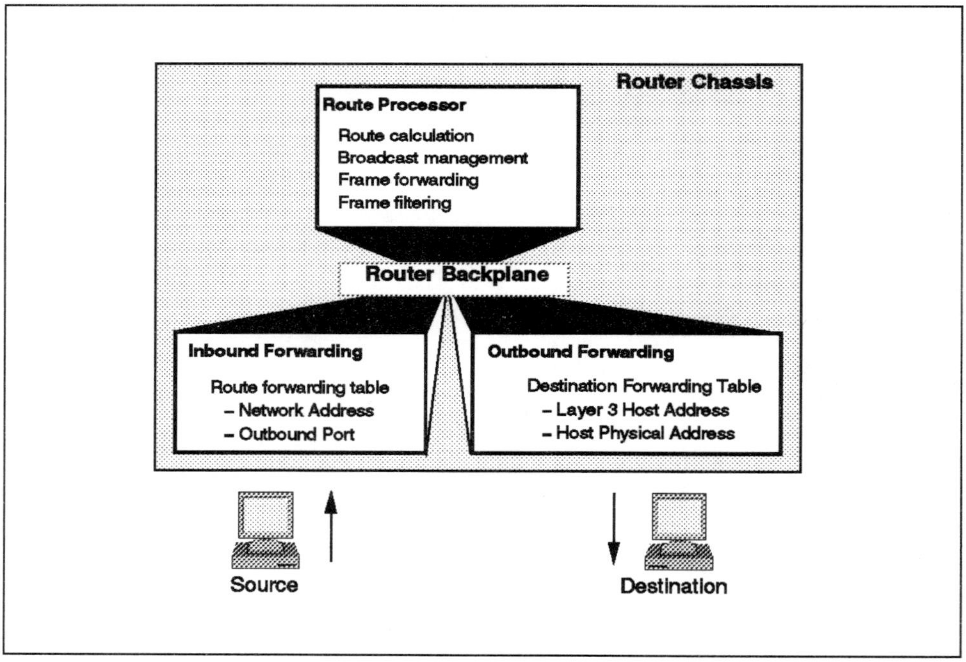

Figure 4. Stand-Alone Router Internals

This is not to say that the *routing function* will no longer be needed, or even that all routers will disappear. Indeed, router functions will remain critical to network operation. Rather, the goal is to expedite movement of information without forcing users to abandon router-dependent protocols. Ideally, one would push the forwarding function of routing all the way out to the periphery of the network, leaving a protocol-independent network core that would be focused on providing reliable, high-performance standards-based ATM connections. Ultimately, this move will become a necessity if the requirements discussed in 1.2, "The Challenges of High-Speed Networking" on page 2 are to be fullfilled.

This requirement has led to the development of distributed routing. If we first consider the internals of today's stand-alone router (see Figure 4 on page 6), it consists of the following:

- Route processor
- Physical interfaces

The route processor handles route calculation, broadcast management, frame forwarding and filtering. The physical interfaces provide frame forwarding and filtering. Since this function is all contained within a single platform, the finite number of slots and ports, as well as the bus speed, all serve to limit the router's flexibility to support microsegmentation. Moreover, there is no support for quality of service (QOS) or distributed workgroups in the backbone. On the plus side, routers are very effective at providing address resolution, security, and broadcast control.

The move to a pure switching model for workgroups and the backbone is a giant step forward in that both servers and end stations can benefit from QOS support and much higher speeds that can be scalable according to need. In addition, virtual LANs can be enabled. The single weakness in this configuration is the lack of multiprotocol support. Hence, a solution that borrows from the router model, but avoids its inherent limitations is required.

Distributed routing is generally viewed as the solution to the multiprotocol dilemma. In general, distributed routing separates the routing function into two parts, a route server and a route client, as shown in Figure 5 on page 8.

The route server does route calculation and provides routing table update services to the route client. The route client handles frame forwarding and filtering, potentially provides ATM LAN emulation services, and keeps a cache of the routing tables.

This generic model for distributed routing can be applied to a switched environment consisting of a central ATM backbone with a route server attached to one of the ATM switches. LAN attachment is then provided through LAN switches, which also serve as the route client. Since the client maintains routing tables that are updated by the server, the model is still one of multihop or hop-by-hop (see Figure 6 on page 8) routing between the LAN switches; so performance is still restricted at those points.

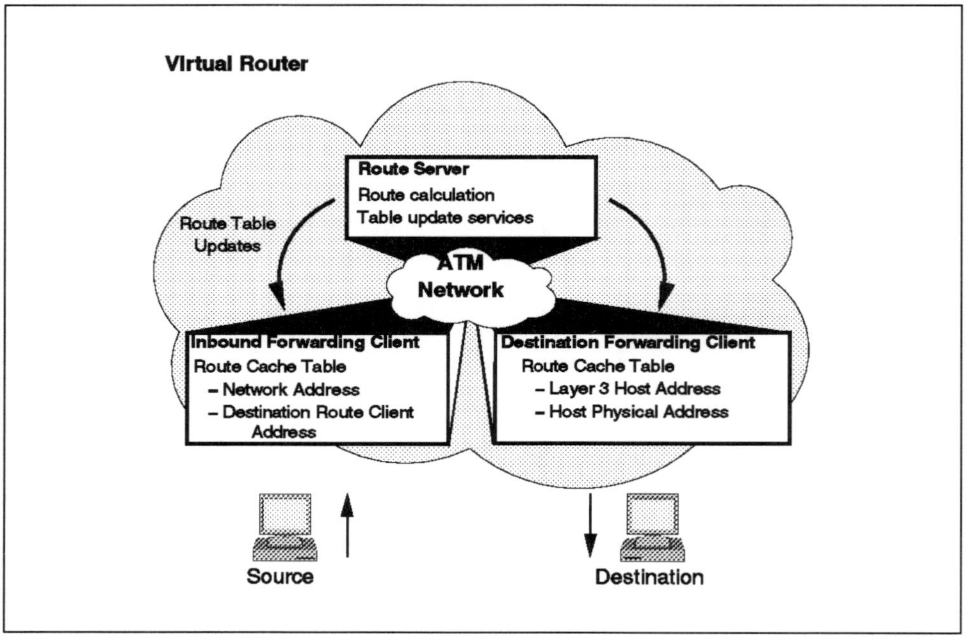

Figure 5. Distributed Routing Model

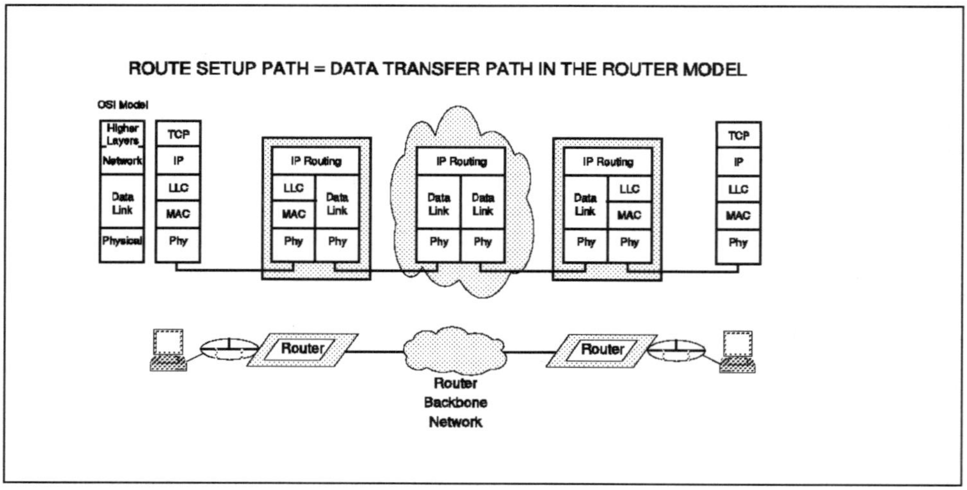

Figure 6. Traditional Hop-by-Hop Multiprotocol Routing

This distributed routing model does have a number of strengths. It provides virtual LAN and virtual subnet support with broadcast control *between* VLANs. It is also scalable in terms of performance (due to multiple processors and no bus limits), number of ports, and redundancy. It does not, however, provide broadcast control *within* VLANs. Without some sort of *intelligent BUS*, broadcast frames still need to be transmitted to every member of a VLAN.

1.2.3 Virtual Networks

The evolving business paradigm has resulted in related functions becoming geographically dispersed. This creates a new entity called a *virtual workgroup*, that is, a group of people who are logically connected, but physically disjoint. This group requires the same level of communications that inspired the creation of the LAN. In other words, even if a workgroup has members spread across the country or around the world, they need to share a single identity to facilitate communication. Reconfiguration of a user node should not be necessary if that user happens to relocate his/her base of operations.

LAN switches provide a fast *virtual media* that eliminates, or at least reduces, congestion, but they historically have little or no capability to segment or subdivide traffic. Dividing the switched LAN into virtual LAN segments or virtual LANs would allow logical communities of interest with relatively high mutual traffic levels to exist as one logical bridged community. A switch-based network could extend those virtual LANs across multiple switches, even multiple locations. This would form a new kind of virtual LAN, which we'll call a *virtual network*. The virtual network is the solution to the problem of the virtual workgroup (see 2.3, "Virtual LANs" on page 60).

1.3 Why ATM?

ATM is a connection-oriented packet-switching technology that uses fixed-size packets, referred to as *cells*, to carry the traffic in the network. ATM embodies various design objectives that include:

- Integration of voice, video, image and data services into a single framework
  - Characterizing ATM as asynchronous indicates that cells may occur at irregular times determined by the nature of the application rather than the framing structure of the transmission system. In effect, ATM has isochronous support *built in*; consequently, ATM can transport voice, data, and video, all on the same circuit.
- Scalability, both in terms of:
  - Distance - A single technology in the local area, campus, and wide area
  - Speed - Currently defined physical layer interfaces vary from 1.5 Mbps up to 622 Mbps

Because of its adaptability, ATM is viewed almost universally as the key technology for the future in carrier networks, in private wide area networks, and on the campus as the multimedia LAN. One common misconception, though, is to confuse the *technology* with the *services* that will be offered in the broadband network of the future. The types of services available in ATM networks include:

- Bandwidth on demand
- Guaranteed service levels
- Point-to-point and point-to-multipoint connections
- Constant as well as variable bit rate services
- Connection-oriented or connectionless application services

To understand the ATM philosophy, first consider one view of networking history.

In the past, networks were either circuit-oriented or packet-oriented. Circuit-oriented services are traditionally preferred for delay-sensitive (isochronous) applications such as telephony, wheras packet-oriented services are preferred for delay-tolerant applications such as data transmission.

With circuit-oriented services, the entire bandwidth of a communications link is dedicated to a specific traffic connection (for example, a telephone call). While this dedicated connection guarantees predictable service, it is also characterized by an inefficient use of the bandwidth. For example, in the case of the telephone call, when no one is talking, the bandwidth lies unused.

Packet switching offers considerably more versatility than the circuit-oriented alternative. The bandwidth in a particular link is not dedicated to a single connection; therefore, multiple sources can utilize this link by having their traffic *packets* multiplexed over it. This minimizes the idle time on the link, but also introduces an element of randomness; the packets are of varying lengths and they arrive at random times. If, for example, a voice or video connection were to be multiplexed with a data transmission, large data packets could interrupt the flow of the voice or video traffic, resulting in a breakup or *jitter*. That is obviously not acceptable to this isochronous form of traffic. Nevertheless, packet switching optimizes the use of bandwidth, which makes packet switching the more economical choice for data-oriented transmission.

ATM combines the best of both worlds to provide the predictability of circuit switching with the flexibility of packet switching. To accomplish this, ATM uses simplified packet-switching techniques, but segments the packets into 53-byte cells, each with 48 bytes of user data and a 5-byte header to be switched into multiple virtual channels and paths. Since the switching can be done in hardware, as described later, overhead inefficiencies can be held to a minimum. Isochronous traffic and data cells can be interleaved in a way that allows quality-of-service guarantees.

1.3.1 ATM Concepts

The key concepts of ATM are as follows:

Cells

All information (voice, image, video, data, etc.) is transported through the network in very short (48 data bytes plus a 5-byte header) blocks called *cells*.

Routing

Information flow is along paths (called *virtual channels*) set up as a series of pointers through the network. The cell header contains an identifier that links the cell to the correct path for it to take towards its destination.

Cells on a particular virtual channel always follow the same path through the network and are delivered to the destination in the same order in which they were received.

Hardware-Based Switching

ATM is designed so that simple hardware-based logic elements may be employed at each node to perform the switching. On a link of 1 Gbps, a new cell arrives and a cell is transmitted every .43 microseconds. There is not a lot of time to decide what to do with an arriving packet.

Adaptation

At the edges of the network, user data frames are broken up into cells. Continuous data streams, such as voice and video, are assembled into cells. At the destination side of the network, the user data frames are reconstructed from the received cells and returned to the end user in the form (data frames, etc.) that they were delivered to the network. This adaptation function is considered part of the network but is a higher-layer function from the transport of cells.

Error Control

The ATM cell switching network only checks cell headers for errors and simply discards errored cells.

The adaptation function is external to the switching network and depends somewhat on the type of traffic, but for data traffic it usually checks for errors in data frames received, and if one is found, then it discards the whole frame.

At no time does the ATM network attempt to recover from errors by the retransmission of information. This function is up to the end-user devices and depends on the type of traffic being carried.

Flow Controls

In its original conception, an ATM network had no internal flow controls of any kind. The required processing logic was deemed to be too complex to be accommodated at the speeds involved. Instead, ATM was envisaged to use a set of input rate controls that limit the rate of traffic delivered to the network.

Since its original conception, ATM has changed somewhat. Flow and congestion controls are back on the agenda but in a very different form from those that exist in traditional networks.

Congestion Control

There is only one thing an ATM network can do when a link or node becomes congested. Cells are discarded until the problem has been relieved. Some (lower-priority) cells can be marked such that they are the first to be discarded in the case of congestion.

Connection endpoints are *not notified* when cells are discarded. It is up to the adaptation function or higher-layer protocols to detect and recover from the loss of cells (if necessary and possible).

1.3.1.1 Cell Switching

The concept of cell switching can be thought of as either a high-performance form of packet switching or as a form of statistical multiplexing performed on fixed-length blocks of data.

A cell is really not too different from a packet. A block of user data is broken up into packets or cells for transmission through the network. But there are significant differences between cell-based networks and packet networks.

1. A cell is fixed in length. In packet networks, the packet size is a fixed maximum (for a given connection), but individual packets may always be shorter than the maximum. In a cell-based network, cells are a fixed length, no more and no less.
2. Cells tend to be a lot shorter than packets. This is really a compromise over requirements. In the early days of X.25, many of the designers wanted a packet size of 32 bytes so that voice could be handled properly. However, the shorter the packet size the more network overhead there is in sending a given quantity of data over a wide area network. To efficiently handle data, packets should be longer. (In X.25 the default packet size supported by all networks is 128 bytes.)
3. Cell-based networks do *not* use link-level error recoveries. In some networks there is an error checking mechanism that allows the network to throw away cells in error. In others, such as in ATM (described later), only the header field is checked for errors, and it is left to a higher-layer protocol to provide a checking mechanism for the data portion of the cell, if needed by the application.

Packets

The term *packet* has many different meanings and shades of meaning depending on the context in which it is used. In recent years the term has become linked to the CCITT recommendation X.25, which specifies a data network interface. In this context a packet is a fixed maximum length (default 128 bytes) and is preceded by a packet level header that determines its routing within the network.

In the late 1960s, the term packet came into being to denote a network in which the switching nodes stored the messages being processed in main storage instead of on magnetic disk. In the early days, a message switch stored received data on disk before sending it on towards its destination.

In a generic sense, packet is often used to mean any short block of data that is part of a larger logical block.

In ATM the word packet is *defined* as a unit of data that is switched at the network layer (layer 3) of the ISO model.

Figure 7 on page 14 shows a sequence of cells from different connections being transmitted on a link. This should be contrasted with the TDM (time division multiplexing) technique where capacity is allocated on a time slot basis regardless of whether there is data to send for that connection. Cell-based networks are envisaged as ones that use extremely fast and efficient hardware-based switching nodes to give very high throughput, that is, millions of cells per second. These networks are designed to operate over very low error rate, very high-speed digital (preferably optical) links.

The reasons for using this architecture are:

- If we use very short fixed-length cells, then it simplifies (and therefore speeds up) the switching hardware needed in nodal switches.
- The smaller the cells can be made, the shorter the transit delay through a network consisting of multiple nodes.
- The statistical principle of large numbers means that a very uniform network transit delay with low variance can be anticipated with the cell approach.
- Intermediate queues within switching nodes contain only cells of the same length. This reduces the variation in network transit delays due to irregular-length data blocks (which take irregular lengths of time to transmit) in the queues.
- When an error occurs on a link (whether it is an access link or a link within the network itself) then there is less data to retransmit. This could be the case, but with ATM if a cell is lost for any reason (such as link error or discard due to congestion),

the whole frame of which it is part must be retransmitted. In a well designed network using optical links (low error rates), this should not bother us too much.

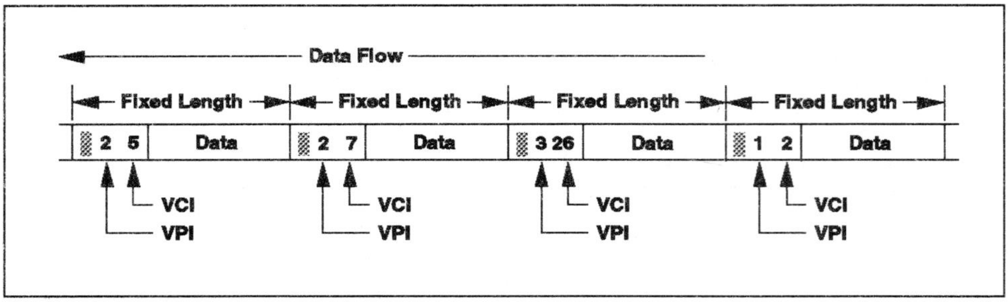

Figure 7. *Cell Multiplexing on a Link*

Note: Cells belonging to different logical connections (identified by the VPI and VCI) are transmitted one after the other on the link. This is not a new concept in the data switching world, but it is quite different from the fixed multiplexing techniques used in the TDM approach. The reasons that cell switching has not been popular in the past are as follows:

- High error rate analog links potentially cause too high an error rate to allow end-to-end recovery. In most cases, link-level error recovery is needed.
- Processing time (load due to processing instructions), both in the network nodes themselves and in the attaching equipment, is greatly increased. Most software-driven data switching equipment takes about the same amount of processor time to switch a block regardless of the length of that block. (This is not exactly true due to the effects of I/O interference, but that is usually small.) For example, if a 1 KB block is broken up into eight 128-byte packets, then the load on the network switching nodes is multiplied by eight.

  The use of hardware logic for routing in cell-based switches minimizes this effect. However, even in a system where hardware-based routing is used, there is significant overhead in the end-user equipment needed for breaking the user data block up into cells and in doing adaptation layer processing.
- The additional bandwidth taken up by headers. The network routing header is necessary, but it takes link capacity to transmit and is an overhead. In the previous example, where a single block is broken into eight smaller packets, then we have multiplied the overhead by 8. This is why packet and cell networks are designed to use very short headers. The need for very short headers is perhaps the primary reason for using connection-oriented protocols.

  In the days when a 2400-bps link was considered fast, this was a significant overhead. In 1996 the cost of link capacity (or bandwidth) is being reduced daily so this overhead is no longer considered significant.

- Hardware technology had not progressed to the point where total hardware switching was economically feasible. (It has been technically possible for some years.)
- In the older technology, end-to-end error recovery processing added a very significant cost to the attaching equipment (significantly more storage and instruction cycles required). This is needed with cell networks today, but hardware cost has become much less and is no longer a problem.
- If the network is designed to do its congestion control by discarding data, and if error recovery is done by the retransmission of whole blocks, then there is a very nasty multiplier effect that can severely impact network performance.

The cell technique is intended to provide the efficiencies inherent in packet switching without the drawbacks that this technique has had in the past. Because cells are small and uniform in size, it is thought that a uniform transit delay can be provided through quite a large network and that this can be short enough for high-quality voice operation.

1.3.1.2 Quality of Service

In addition to the basic functionality and capacity requirements for a service, there are many other requirements that characterize the *quality of service* that a network must provide. This section describes the common quality-of-service requirements associated with the provision of high-speed telecommunications services using networks of ATM switching components.

Delay and Delay Jitter: There is always some form of delay associated with providing a service using networks of ATM switching equipment. Real-time services, like voice and video, are sensitive to delay; that is, there is a quality-of-service requirement that the delay associated with the service is:

- Limited, that is, less than some maximum value
- Predictable, that is, the variation in delay is less than some maximum value

The amount by which the delay associated with a service varies is also known as *delay jitter*. For some types of services, a considerable delay may be acceptable provided that it is predictable, that is, the delay jitter is kept small. For other services both the delay and the delay jitter need to be kept small.

Delay and delay jitter are illustrated in Figure 8 on page 16.

To better understand how delay and delay jitter quality-of-service requirements can be satisfied, it is useful to consider the various sources of delay that occur when providing a service using networks of switching equipment.

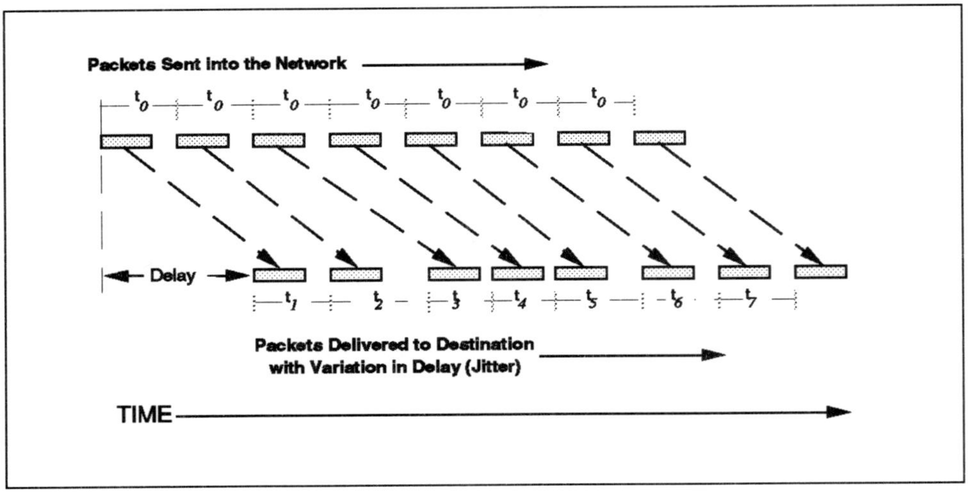

Figure 8. Delay and Delay Jitter

Packetization Delay

When providing services using packet-switched networks[2] such as those based on ATM switching equipment, there is a requirement to packetize information where it enters the network, if it is not already in a packetized form.

Packetization introduces a delay equal to the number of bits in each packet divided by the access speed of the service. Large packets introduce larger packetization delays. Low-speed services also experience larger packetization delays.

Packetization delay is generally predictable; that is, it does not contribute to delay jitter.

Assume user A in Figure 9 on page 17 has a block of 1024 bytes to send through a 3-node network to user B. Assume also that the link speeds are the same, the nodes are infinitely fast, there is zero propagation delay, and that there is no other traffic.

- User A sends to node 1 and takes (for our discussion purposes) 4 units of time.
- Node 1 sends to node 2 also taking 4 units of time.
- Node 2 sends to node 3.

2 The term *packet* is normally used for variable-length packets, whereas in ATM the term *cell* is used for the fixed-length (53-byte) packet that is used to transport data through an ATM network. This distinction is irrelevant for the discussion in this section.

- And so on until the message arrives at user B.

The total time taken has been 4 x 4 units = 16 units of time.

Now, if the 1024-byte block is broken up into four 256-byte packets, then the following scenario will occur:

- User A sends the first packet to node 1, taking 1 unit of time.
- Node 1 sends this packet to node 2, but while this is happening, user A is sending packet 2 to node 1.
- While user A is sending the third packet, node 1 is sending a packet to node 2 and node 2 is sending a packet to node 3.
- This happens through the network until the last packet arrives at user B.

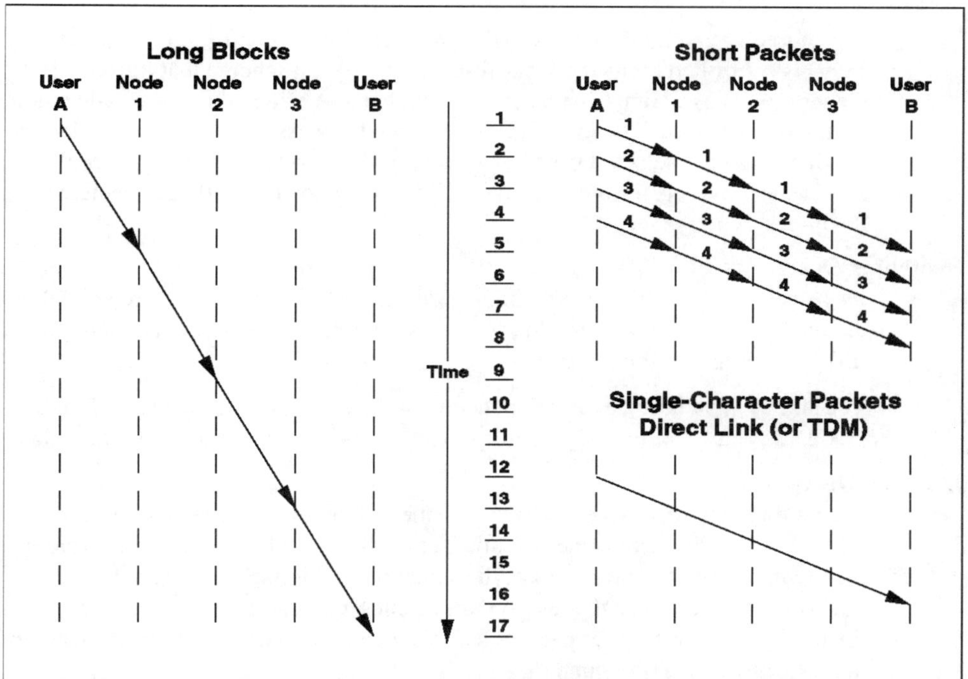

Figure 9. Effect of Packetization on Transit Time through a 3-Node Network

It is obvious from the diagram that sending the message as small packets has reduced the network transit time to 7 units compared with the 16 units needed without packetization. This is due to the effect of overlapping the sending of parts of the message through the network.

The section of the figure headed Direct Link refers to what happens in the limit as we keep reducing the packet size. When we finally get to a packet size of one character, we have a network throughput equivalent to a direct link or TDM operation of the nodes. Transit time in this case is 4 time units.

Admission Delay

Admission delay is a delay deliberately introduced where a service enters the network to satisfy the needs of a network resource allocation scheme. The amount of resources required to support a particular service can be reduced in some circumstances by introducing a small delay to packets associated with that service.

Admission delay is also known as *smoothing* delay or *shaping* delay as it is often introduced to smooth or shape traffic associated with variable bit rate (VBR) services.

The amount of admission delay and its predicatability (contribution to delay jitter) is controlled by the network resource allocation scheme that introduces it. Generally, a large admission delay can enable more efficient resource allocation within the network. Therefore, the admission delay associated with a particular service is usually selected based on the quality-of-service requirement for the service, estimations for other delays and the desire to efficiently utilize network resources.

Switching Delay

As packets traverse transit switching nodes in an ATM network, there are delays associated with the switching fabric used to transfer packets from an incoming link to an outgoing link.

The size of switching delay is dependent on the design of the switching fabric. Good switching fabric designs will provide low switching delay and jitter delay.

Queuing Delay

Where a variety of services utilize the same link or any other shared resource within a network, there is the potential for contention. That is, several services may require use of the resource at the same time. For packet-switched networks, such as those that use ATM switches, contention is resolved by queuing packets whenever packets associated with two or more services want to access a resource at the same time.

When packets are queued (mostly because the outgoing link is busy), this introduces a delay. The size of this delay is quite difficult to predict and depends on many factors, such as the types of services being provided by the network and the traffic characteristics of those services. However, in general, the average queuing delay will become larger as the utilization of network resources increases.

Queuing delay can be minimized through the careful allocation of network resources. A major challenge in the implementation of networks of ATM switching equipment is the allocation of network resources so that queuing delays can be managed effectively.

The predictability of queuing delay (and therefore its contribution to delay jitter) is dependent on the types of services being provided by a network. When providing variable bit rate (VBR) services, there can be considerable variation in queuing delay and, therefore, considerable contribution to delay jitter.

Transmission Delay

This delay occurs because it takes a finite time to transmit a bit of information over a link. This time is directly related to the speed at which the link is operated.

The impact of transmission delay depends on the packet size used to transport services across a network. Larger packets will experience larger transmission delays, while smaller packets will experience smaller transmission delays. For example, on a link operated at 2 Mbps, a 53-byte packet (424 bits) will incur a transmission delay of 0.212 ms. On the same link, a 1000-byte packet (8000 bits) will incur a transmission delay of 4 ms.

Transmission delay can be kept small by either using small packets or high link speeds. Traditionally, when link speeds were low, transmission delay was significant. However, as link speeds increase over the next few years, the impact of transmission delay will become less significant.

Transmission delay is generally predictable; that is, it does not contribute to delay jitter.

Propagation Delay

This delay occurs because the electromagnetic waves used to transport digital signals can only propagate down a wire or an optical fiber or through free space at a speed that is less than or equal to the speed of light (300000 km per second). In fact, the actual speed for propagation over wire or optical fiber is less than the speed of light (usually in the range 60% to 90%) and depends on a characteristic of the wire or fiber called the *velocity factor*.

Propagation delay becomes significant where there are long-distance links in a network. For example, a 5000-km transcontinental link will have a minimum propagation delay of 16.7 ms. The actual delay will be much longer depending on the velocity factor of the wire or optical fiber providing the link.

Propagation delay is generally predictable; that is, it does not contribute to delay jitter.

Playout Delay

Playout delay is a delay deliberately introduced where a service exits the network to smooth out the effects of delay jitter. It is generally used when providing services that are jitter-sensitive, such as voice services.

Delay jitter can be reduced if all packets associated with a service are buffered before they exit the network and *played out* of the buffer at a constant rate.

Playout delay is also known as *reassembly* delay as it is often introduced when reassembling packets associated with a CBR service to recreate a continuous bit stream.

The amount of playout delay is selected to equal the maximum delay jitter that should be cancelled.

1.4 ATM in the Wide Area Network

In the wide area network environment, ATM offers a number of significant benefits:

Integration of Services

The number of specialized services offered by carriers around the world has proliferated in the last few years. This is seen as a direct response to the proliferating needs of users. One of the benefits of ATM is the ability to satisfy most user requirements with a single service and to reduce proliferation of new kinds of networks.

Lower Equipment and Network Cost

It is widely believed that ATM switching equipment will cost significantly less than time-division multiplexing (TDM) equipment to do the same job.

Appropriate Technology to the High-Speed Environment

ATM offers a technology that can deliver service at the very high speeds now becoming available and being demanded by users.

There are two quite distinct environments here:

1. The carrier environment, where ATM is provided as a service to the end user.
2. The private network environment, where a large organization purchases lines from a carrier (or installs them itself) and builds a private ATM network.

The major problem with the wide area environment is government regulation. In most countries, governments regulate the detailed technical characteristics of anything and everything that connects to a public communications network. This is often called *homologation.* Part of the homologation process requires protocol testing, which involves the detailed and exhaustive checking of the communication protocol as it is performed by the device being tested.

In years gone by, homologation was an essential task (malfunctioning or badly designed end-user equipment could not only disrupt the network's operation, but in some cases actually cause damage to network equipment). Those days have gone forever. Today's communication switches (telephone exchanges, ATM switches, etc.) are computers. They have quite sufficient logical ability to rigorously ensure that attaching equipment cannot disrupt or damage the network. But governments (on the advice of the people who do the testing) continue to insist that testing is needed.

Protocol testing is an extremely expensive and very slow task. The requirement for protocol testing has very significantly retarded the growth of (regular) ISDN systems around the world. ATM in the WAN environment will develop only very slowly if this situation continues.

1.5 ATM in the LAN Environment

Many people believe that it will be the LAN environment in which ATM gets its first significant usage. There are good reasons:

Users Need a Higher-Capacity LAN System

Existing shared-media LANs were designed and developed in an environment where communications bandwidth (bits per second) was almost free. Compared to the internal speeds (and the possible I/O rates) of early personal computers, the LAN was very nearly infinite.

Most wide area network architectures were developed with the overriding objective of saving the cost of bandwidth. This was done by spending money on devices (in hardware cycles, memory, and software complexity) to optimize (minimize) the cost of bandwidth. When people came to build LANs, the use of existing WAN networking systems seemed inappropriate. Why spend money to optimize a free resource? Thus LAN network protocols save cost by using extra bandwidth. This is *not* inefficient; this is a valid and reasonable adjustment to a changed cost environment (even though it did not take account of the future need to interconnect LANs over the wide area).

As personal computers and workstations have increased in capability, there has been an attendant rise in the demand for LAN capacity. In the short term, this can be solved by restructuring large LANs into smaller ones, bridging, routing, and the like, but the improvement in workstation capability and the increase in bandwidth demand does not look likely to slow down, especially when going into multimedia applications. A faster (higher-throughput) LAN system is needed.

ATM Cost/Performance in the LAN Environment

Looking at Figure 10 on page 22 and Figure 11 on page 23, the performance advantage of a switched system over a shared-medium system is easily seen. In

Figure 10 on page 22, let us assume that we have 50 devices connected to a 100-Mbps FDDI LAN. This is a shared-medium system. Only one station may transmit data at any one time. This means that the total *potential*[3] network throughput is 100 Mbps.

In Figure 11 on page 23, we have the same devices connected through a switch (let us assume a link speed of 100 Mbps). Each device is capable of sending at full media speed.[4] The total *potential* network throughput here is not 100 Mbps but 50 x 100 Mbps, or 5000 Mbps.

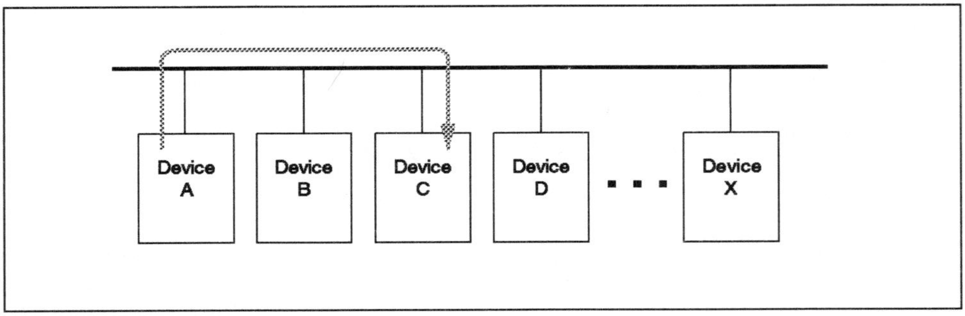

Figure 10. *Connection to Shared Medium*

The difference is in the cost. The adapter cards for 100 terminals connected to an FDDI LAN (only 100 Mbps total throughput) will cost about $1,000 each or a total of $100,000. 10-Mbps ATM adapter cards (if they existed) could cost perhaps $200 each[5] for a total of $20,000.

Of course, with ATM you need a switch, but with FDDI you need a hub. It is expected that ATM switches will cost only marginally more than FDDI hubs. Both systems need a network management workstation.

In the next example, the ATM system has a maximum potential throughput of 10 x 100 Mbps = 1 Gbps. The FDDI LAN still has a maximum throughput according to the media speed (100 Mbps). The net is that an ATM system will deliver much greater throughput than a shared-medium LAN system for a significantly lower cost.

3 In reality, of course, quite a bit less.

4 In this situation, too, throughput is limited by statistical considerations.

5 The cost is about the same as an Ethernet adapter *plus* some hardware to convert data frames into cells and back again.

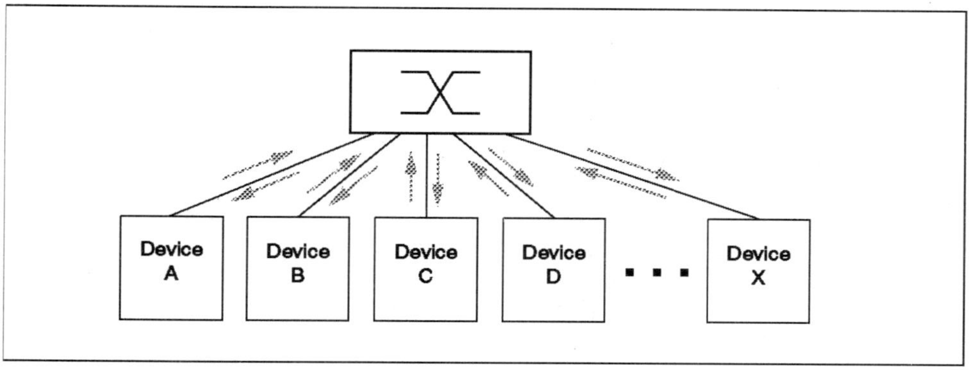

***Figure 11.** Connection to a Switch*

This cost equation also applies to *switched* LANs. The previous discussion could apply almost equally to switched Ethernet as to ATM, at least on a cost basis. However, ATM systems have significant advantages over switched LAN systems:

1. Switched LAN systems don't scale up very well at all. It is difficult to build large networks just using switched LANs. You tend to need routers to interconnect segments of the network. This adds significantly to the cost and cripples the throughput.
2. Switched LAN systems need to use bridges or routers to connect across the wide area. Adding a WAN protocol to connect LANs across a WAN is not an efficient way to connect LANs.
3. Switched LANs don't handle isochronous traffic (video, voice, multimedia) very well, and they do not scale well.

Security

On a shared-medium LAN all of the data traffic passes through every workstation. While some types of LANs have a higher level of security than others, it is true that data security in the LAN environment leaves a lot to be desired.

In ATM, a single end-user station only receives data intended for that station. Thus, users cannot *listen in* to what is going on in the rest of the network.

Asymmetric Bandwidth

In a LAN environment some users will require much higher data throughput than others. Servers, in particular, need much higher throughput than any of their clients. In a shared-medium system, all these devices must have adapters that run at the same speed (the speed of the LAN).

In ATM, individual stations can have connections at speeds appropriate to their capability. A speed of 25 Mbps is faster than any current (ISA bus) PC. Many workstations can easily handle a throughput of 50 Mbps and specialized servers are being built to handle total rates of a few hundred Mbps. This ultimately saves cost, both in the workstations themselves and in the network switching equipment.

ATM Supports Isochronous (Timing-Dependent) Traffic

Perhaps the most discussed near-term application is for multimedia of one form or another on workstations. This usually requires video and sound as well as traditional data. The video and sound components require a system that can deliver data at a timed rate (or can adjust to one). A number of proposed LAN architectures to do this are being explored (isochronous Ethernet, FDDI-II), but ATM offers the potential to allow these applications in an integrated way.

Reduced Need for Routers and Bridges

In an ATM environment there is still a need for routers and bridges, but in some network configurations, the need is significantly less, for example, where a number of different LAN protocols are used by groups of devices on a set of linked LANs. In an ATM system a number of separate virtual LANs can be set up (one for each protocol) and the need for a router is removed. The bridging function between dispersed locations is performed by ATM without the need for bridges as such. Bridges are only needed to connect between ATM-attached workstations and workstations attached to existing LANs.

1.6 ATM as a LAN Transport Mechanism

If we consider the standard 7-layer OSI reference model (see Appendix A, "Protocol Stack Reference" on page 193), legacy LANs and ATM devices may be interconnected at various different levels (the choice of level determining the degree of function that can be achieved, as shown in Figure 12 on page 25).

To enable internetworking between these devices, we require that the applications residing on ATM devices share a common interface. There are essentially two types of interface available as follows:

- High-level application programming interfaces (APIs), for example, APPC and sockets, provide transport layer services to users.
- Low-level interfaces, for example, NDIS and ODI, provide users with access to the MAC services of legacy LANs.

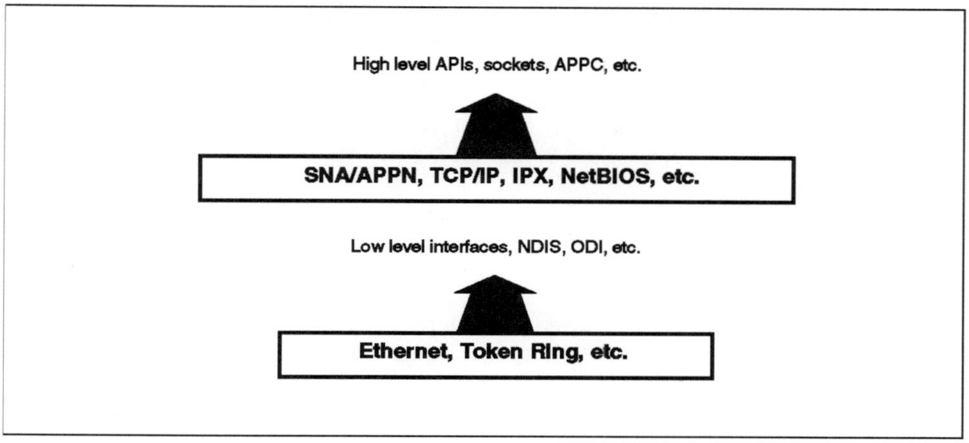

Figure 12. *Programming Interfaces*

If we consider the high-level API, we could use the ATM network as a new link layer technology and modify the existing network layer 3 protocols as required. This method has been used by the IETF in the definition for Classical IP over ATM (RFC 1577). This approach is attractive where the dominant communication protocol is IP. However, many other protocol stacks are quite widespread: SNA/APPN, IPX, NetBIOS to name a few. Porting these protocol stacks individually would take considerable effort.

To avoid this portability issue, one could consider supporting the high-level APIs directly over ATM. With this solution a transport protocol must be added above the ATM protocol stack, and connectivity to legacy LANs is performed using gateways implementing internetworking at the transport layer. Candidates for an ATM transport protocol may be an existing one (for example, TCP) or a new one designed specifically for cell-based ATM networks. The main concern with this approach is that the definition of a transport protocol for ATM may become a major standardization task.

These issues concerning the support for high-level APIs could be avoided if we consider the low-level interface, which gives access to the MAC services of the legacy LAN. With this approach, a layer is also added above the protocol stack. However, instead of providing transport services, this new layer emulates the MAC services of a legacy LAN, (for example, Ethernet or token-ring). Because the MAC layer services of an existing LAN are emulated, any higher-layer communication software (for example, TCP/IP and SNA/APPN) written for the legacy LANs can be reused without change. This is the main appeal of the ATM Forum *LAN emulation* method. However, this is also the weakness of the method, because it provides the services of a legacy LAN and not those of an ATM network.

Depending on the functions required and levels of complexity that we are willing to implement for any given solution, there are three different models for transport over ATM:

Subnet

For example, LAN Emulation Version 1, as shown in Figure 13.

- One subnet per ATM LAN
- Intrasubnet Communications via ATM Virtual Circuits
- Routers Interconnect ATM and Legacy Subnets
- Intersubnet Communications through Routers
  - ATM VCs initiate/terminate at router.
  - QOS function does not cross subnet boundaries.
- Traditional Internet model - Well understood, easy to implement

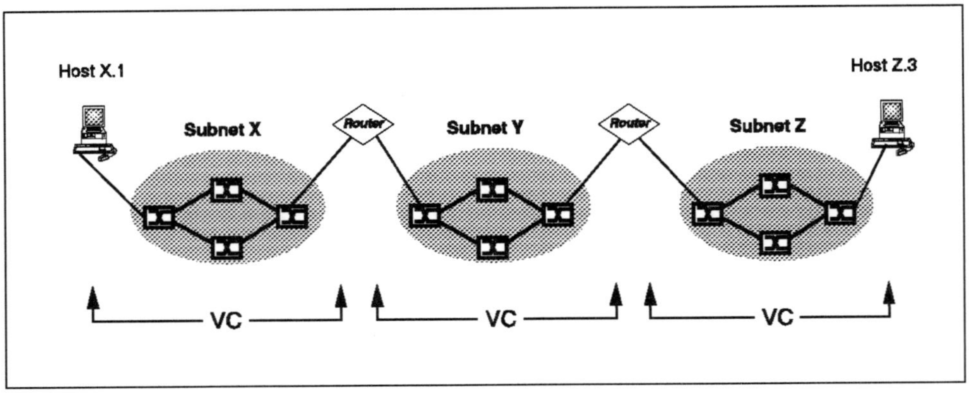

Figure 13. Subnet Model

Integrated

For example, I-PNNI, as shown in Figure 14 on page 27.

- Routers and switches share single routing protocol
  - Single topology database and route computation.
  - Best path computed based on end-to-end topology metrics and reachability.
- Cut-through possible between subnets
  - Routers cannot forward SVC requests.
- QOS enabled over ATM network

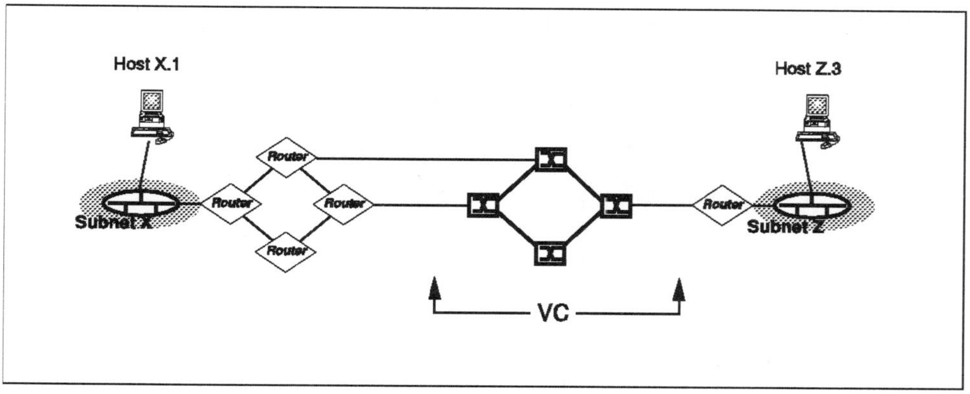

Figure 14. Integrated Model

Cut-Through

For example, MPOA using NHRP for address resolution, as shown in Figure 15.

- Multiple subnets share single physical ATM network
- Protocols used to establish direct SVCs across subnet boundaries
  - Host X.1 establishes direct connection (SVC) across ATM network with Host Z.3.
- Routers Interconnect ATM and legacy subnets
  - Default packet forwarder.
  - Help resolve internetwork and ATM addressing - Forward or respond to queries.
- QOS enables over ATM network

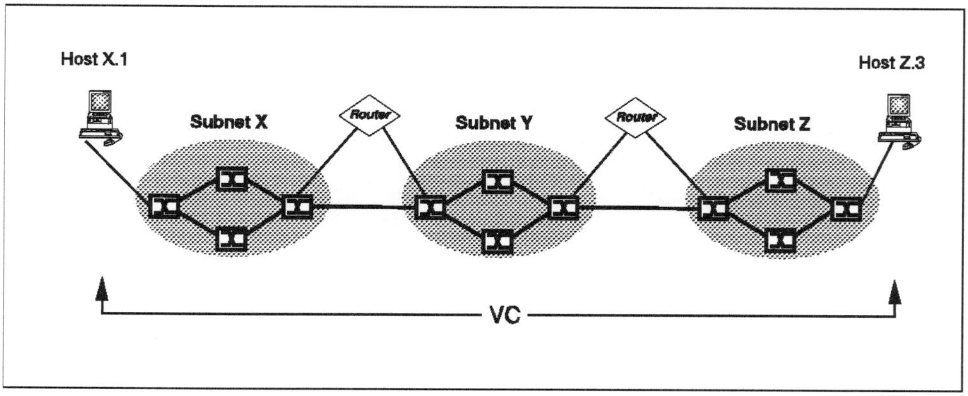

Figure 15. Cut-Through Model

1.7 Legacy Protocols in ATM Networks

In order to understand why legacy protocols are being replaced in ATM networks, the following sections explain some of the basics of address resolution and routing in legacy networks. This section gives a basic overview of various functions, see later chapters for more detailed explanations.

1.7.1 Address Resolution

In legacy link layer protocols, a sending node must know the MAC (media access control) or layer 2 address of the node it wants to send data to before the nodes can communicate. The networking protocol, for example IP, uses a networking address (layer 3) and not the MAC address (layer 2) of a target. Hence, ARP (address resolution protocol) provides the MAC address for a given network address. In IP networks, address resolution is provided by ARP (address resolution protocol). An ARP cache in each node can contain a list of known mappings between MAC and network addresses.

1.7.1.1 Legacy

Figure 16 on page 29 shows the basic ARP process. The source 9.68.1.1 needs to send data to the target 9.68.1.2. Initially, the ARP cache does not contain the MAC address of the target. To obtain the target's MAC address, the source ARPs for 9.68.1.2. This is done by (in the case of Ethernet which is used for this example) transmitting a broadcast frame on the Ethernet. All nodes on the segment receive this frame. As the target has the IP address being searched for, it sends an ARP reply back to 9.68.1.1. The ARP reply contains the MAC address of 9.68.1.2. Source 9.68.1.1 adds the MAC/network address pair to its ARP cache and sends the IP data for node 9.68.1.2 to MAC address 400010001111. In this case, the IP addresses were predefined in each node, and the MAC addresses were either locally or universally administered addresses, known by the node itself.

Although a very basic explanation, it shows that the address of a partner is found by broadcasting to all nodes on the same medium. This is not possible when using a switched network.

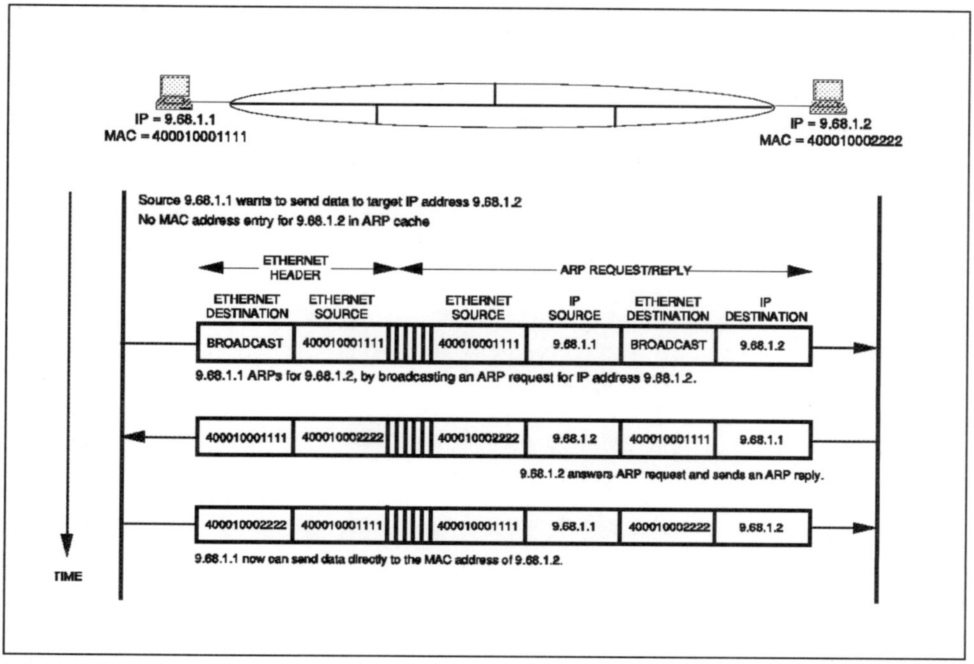

Figure 16. ARP

1.7.1.2 ATM Networks

LAN Emulation

Address resolution in a LANE environment is shown in Figure 17 on page 30. (This diagram leaves out some details, for example, cases where the requested information is already cached, etc.) To overcome the inherent lack of a broadcast function in an ATM environment, the broadcast and unknown server (BUS) function is used. See 2.1.2, "LAN Emulation Components" on page 48 for more information. The IP addresses are defined as usual, and the MAC address is either locally or universally administered. The ATM address of the LES is either preconfigured or supplied by the LECS. The address of the LECS is obtained via the interim local management interface (ILMI), or the well-known LECS ATM address is used.

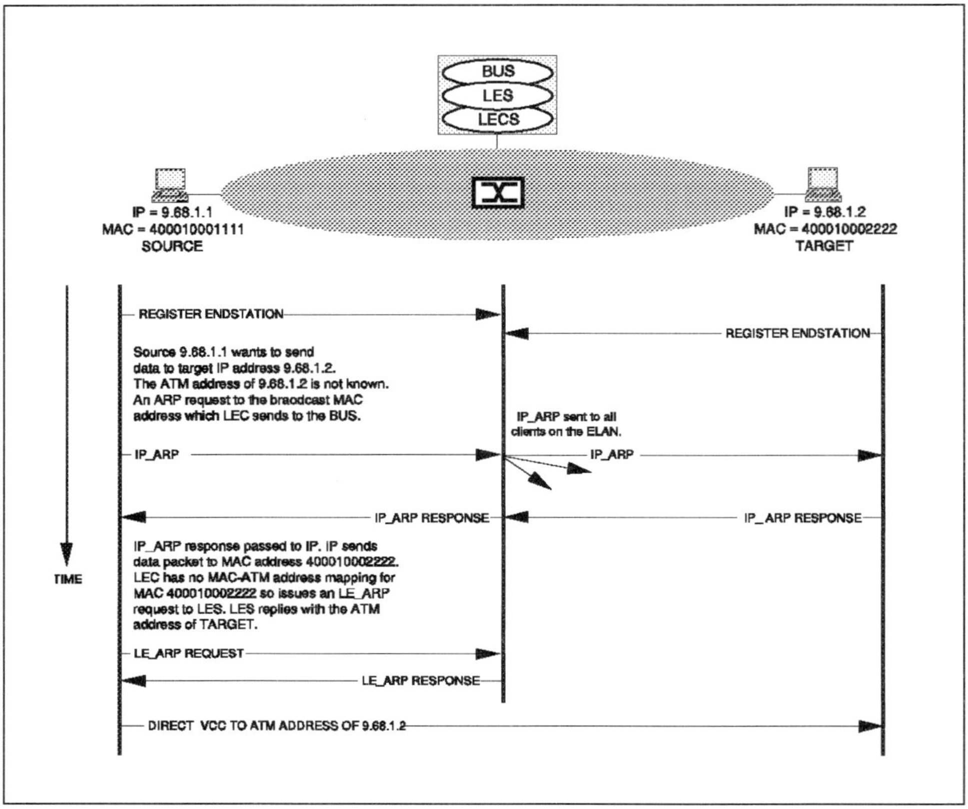

Figure 17. LANE ARP

When source 9.68.1.1 wants to send data to 9.68.1.2, and the MAC address is not known, an ARP request is sent to the broadcast MAC address, usually 'FFFFFFFFFFFF'. The LAN emulation client (LEC) software sends MAC broadcast frames to the BUS. The BUS then transmits the broadcast IP ARP to all nodes on its multicast send VCC or multicast forward VCC. The station with IP address 9.68.1.2 answers this IP ARP with its MAC address, which goes back to the BUS who sends it back to 9.68.1.1. Once the IP layer in 9.68.1.1 has the destination MAC address, it tries to send the data packet to that MAC address. The packet is passed to the LEC that looks for a MAC-to-ATM address mapping for that MAC address. If no mapping can be found, it creates an LE_ARP_REQUEST that is sent over a VCC to the LES. The LES answers with an LE_ARP_RESPONSE that contains the ATM address for that MAC address. The LEC then uses this ATM address to establish a VCC to the target. In addition, the LEC can also add this mapping to its own LE_ARP cache.

LANE with LNNI

The LAN emulation V1 specification does not specify how LE server entities should communicate with other LE server entities. The LAN Emulation V2 specification (which is still under development) includes the LAN emulation network-to-network interface specification LNNI. LNNI describes the interface over which LE server entities (LES, BUS and LECS) will communicate with other LE server entities. It does not describe the protocols that will use the LNNI interface. 2.2, "LAN Emulation Version 2.0" on page 59 gives more details on LNNI.

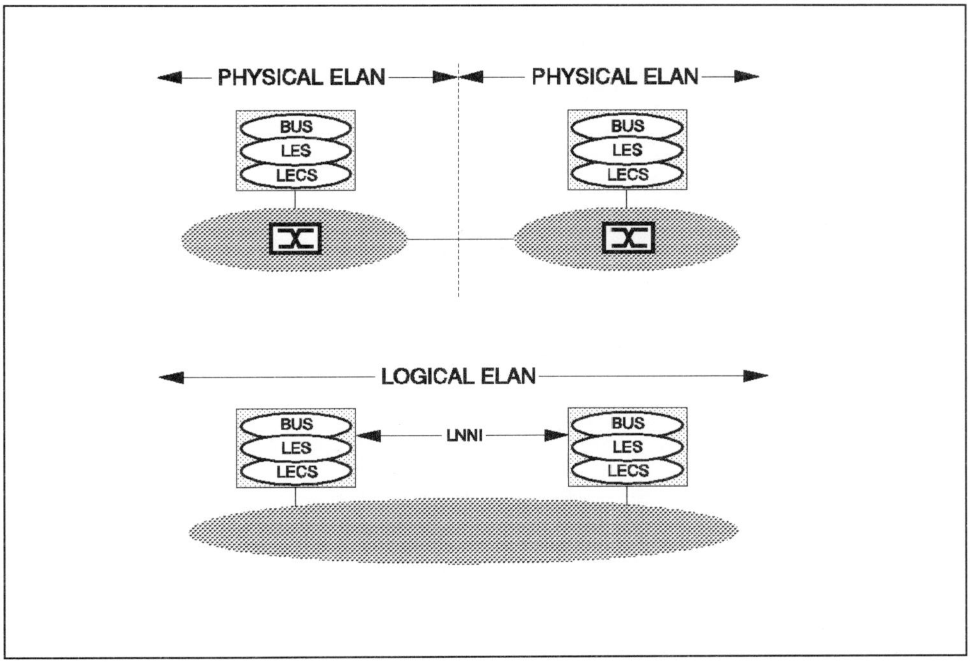

Figure 18. LNNI

LNNI will allow, for example, a BUS that cannot satisfy an LE_ARP_REQUEST to propagate this ARP request to other BUS entities. In general, true broadcast traffic, as opposed to unknown traffic, must be flooded to all BUSs in the network. A spanning tree will be built among the servers to allow routing of such requests. The result will be that at the MAC level, multiple emulated LANs act like one large emulated LAN with distributed server functions. Figure 18 gives an example of this. To all higher layer protocols, the two emulated LANs now act like one single physical LAN. An ARP request on one emulated LAN can be resolved to an ATM address on any of the emulated LANs, as there is ATM connectivity between the ELANs, a VCC can be set up to the partner.

In this document, when we refer to an ELAN, it can mean a single emulated LAN or multiple ELANs connected with LNNI.

ATM Address Resolution Protocol

RFC 1577 defines a protocol to support automatic address resolution of IP addresses in ATM networks, this is known as Classical IP over ATM (IPOA). IPOA introduces the notion of a logical IP subnet (LIS). A LIS is a group of IP nodes that connect to a single ATM network and belong to the same IP subnet. Each LIS has a single ATMARP server. All nodes in a LIS are configured with the ATM address of the ATMARP server to enable address resolution.

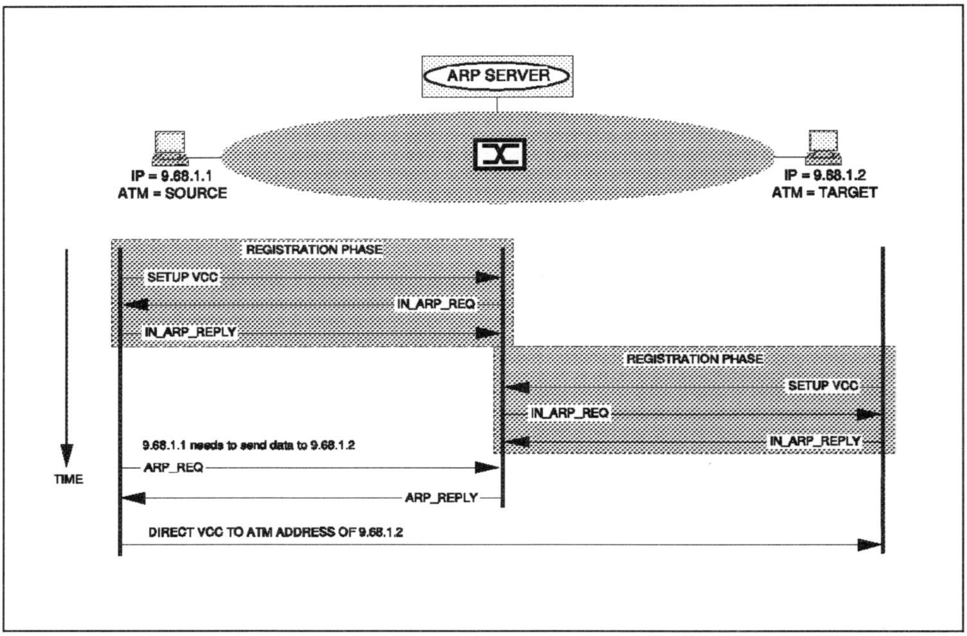

Figure 19. ATM Address Resolution Protocol

When a node is activated, it establishes a VCC to the ATMARP server using the preconfigured address. On receiving a connection from a new client, the server sends an inverse ARP request (IN_ATMARP_REQ) to that client, which requests the node's IP and ATM addresses. The returned address mappings are stored in the server's ATMARP table.

When 9.68.1.1 needs to send data to 9.68.1.2, it sends an ARP request to the server for the ATM address of 9.68.1.2. If known, the server returns an ARP reply containing the ATM address, and a direct VCC can be set up to the target. If not, an ATM_NAK is returned.

In RFC 1577, no MAC addresses are used or needed. Also, the ARP function has been changed to use an ATM VCC instead of a broadcast address. Communication is limited to nodes in the same IP subnet as no routing functions are implemented between ATMARP servers.

Next Hop Resolution Protocol (NHRP)

ATMARP limits address resolution in an ATM network to a single logical IP subnet (LIS). NHRP will be used to resolve the ATM addresses of IP hosts across multiple logical IP subnets. NHRP (see 8 on page 229) introduces the concept of a nonbroadcast multiaccess network (NBMA). An ATM network is an NBMA network, but the scope of NHRP is not limited to ATM networks. See 5.2, "IP Address Resolution in ATM Networks" on page 114 for more details on NHRP.

Figure 20 shows a network with two switches connected to make a single NBMA network. In this diagram, each switch serves a separate LIS. Each LIS must have at least one next hop server (NHS). The two switches are connected by an ATM connection that would enable a direct VCC between the target and source stations. In addition, the logical IP subnets are also connected by a router. Nodes in the network are configured with the address of their NHS.

When a node needs to send data to a node in another LIS, the node issues a next hop resolution request, which is sent to its NHS. As the NHS in LIS #1 does not serve the target node, the request is passed to other next hop servers. The NHS for LIS #2 can resolve the request and returns a next hop resolution reply to the NHS originating the request, which, in turn, returns it to the station that generated the resolution request. Now a direct VCC can be set up between the source and target.

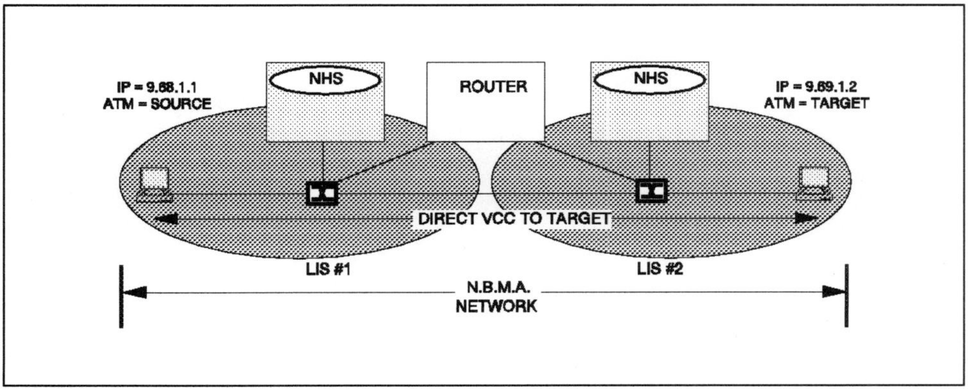

Figure 20. NHRP

NHRP can also return the egress or exit point from an NBMA network that should be used to reach a given target, when that target is outside of the NBMA network. So-called *backdoor* connections (connections between routers that are outside of the NBMA network) can lead to stable routing loops. This problem is not yet solved in NHRP. Indeed it looks as if NHRP will not totally replace layer 3 routing functions, but will only complement them.

NBMA Address Resolution Protocol (NARP)

NARP, or RFC 1735, was an experimental protocol that was created as a discussion document. NHRP evolved out of the work done on NARP.

1.7.2 Routing Protocols

Routers are needed to connect different IP subnets together. In the following sections we discuss how routers and routing protocols work in legacy and ATM networks.

1.7.2.1 Single Router

Legacy Based

To forward IP datagrams between different IP subnets, a router is needed. The router provides a default destination MAC address in the same physical network as the sending station, obtained by the initial ARP from the sending station.

Figure 21 on page 35 shows a simple example where a router is connected to two IP subnets, both IP stations have obtained the MAC address of the router by ARP. If 9.68.1.1 were to attempt address resolution for 9.69.1.2, it could not be successful as 9.69.1.2 does not see the ARP that 9.68.1.1 broadcasts. To overcome this, a default router IP address is defined at each node. The IP address of the default router must be on a MAC address reachable by ARP. When 9.68.1.1 wants to send data to an IP address in another subnet, it ARPs for the MAC address of its default router, and the router returns its MAC address in an ARP response. Once the source node has the MAC address of the default router, packets for the other subnet can be sent to the MAC address of the default router. The router then forwards the packets toward their destination. Once the router has determined which physical adapter the target subnetwork is on, it gets the MAC address of 9.69.1.2 by using ARP in the second subnet.

Routing tables inside the router that are built from the configuration definitions enable it to decide on which port it needs to transmit packets to reach their destination. In this case, the subnetwork connected to a specific port on the router is defined.

What is important here with respect to ATM is that the router must look into the actual IP packet before it can decide on how to pass this packet on towards its destination. ATM cells arriving at the ATM interface of a router must be reassembled into complete IP packets before they can be passed up to the IP layer where the routing decision is made. This is done for every IP packet.

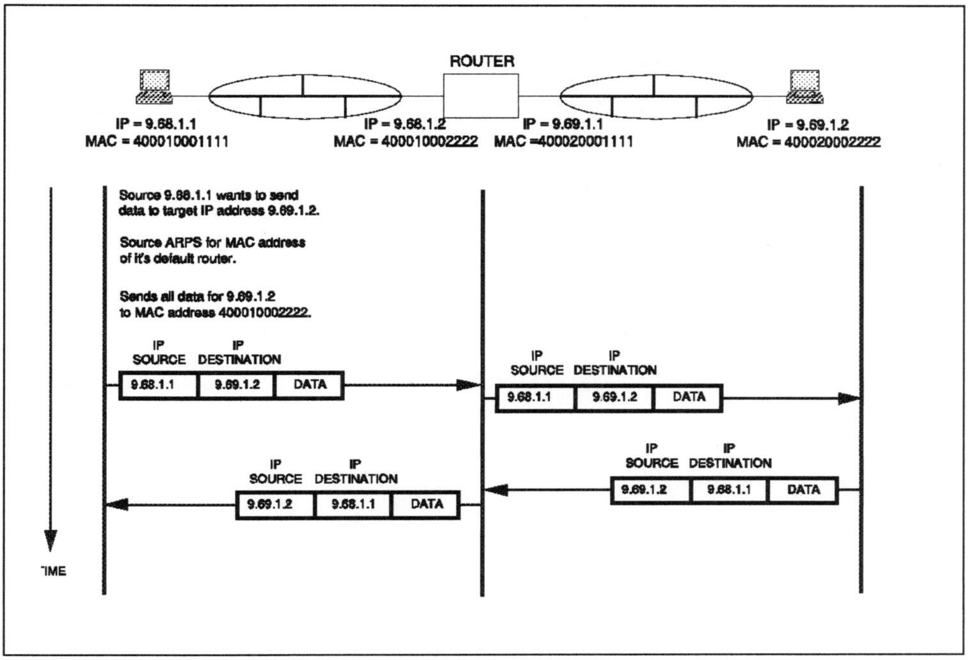

Figure 21. Routing

ATM Based

Emulated LANs

As LANE shields the layer 3 functions from ATM transport specifics, an emulated LAN appears to higher-layer functions as a legacy LAN. The actual physical equipment (single or multiple ATM switches) or number of emulated LANs is totally transparent to the higher-layer protocols. An IP node wishing to communicate with another IP node in a different IP subnet, but on the same ELAN, has two choices. Figure 22 on page 36 shows both methods.

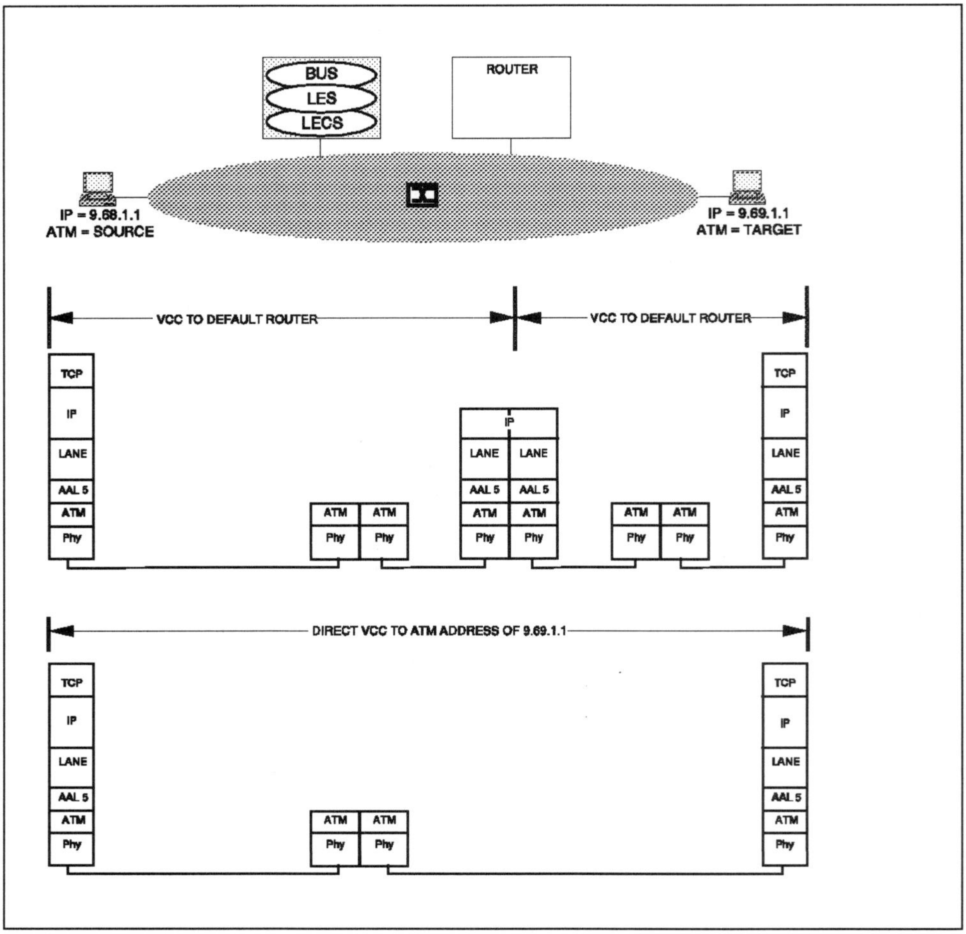

Figure 22. LANE Router

- In the top protocol stack of Figure 22, using LANE functions, each ATM node has a VCC to its default router. This is possible as both ATM nodes are in the same emulated LAN. The router may have one or more connections into the emulated LAN. The protocol stacks shown make it clear that this is not very efficient.

- The lower protocol stack of Figure 22 shows a direct VCC to the target. IP can request the direct VCC if the IP address 9.68.1.1 is defined at the route to network 9.69. While the advantages of the direct VCC are clear, definitions are needed for each other IP subnet in the ELAN.

To enable routing between IP subnets in ELANs, at least one layer 3 router is needed somewhere within the emulated LANs. Figure 23 on page 37 shows an example where a single router serves an ELAN with distributed server components. The router is defined as a default router for all IP nodes in the ELAN. As the server components of the ELAN exchange information over the LNNI interface, an ARP request to find the default router in any part of the ELAN will be successful and return the ATM address of the router. Direct VCCs can then be established directly to the router, even if it on a different part of the ELAN. This allows at least a one-hop route between all nodes in the connected ELANs. There are performance implications to this though; cells containing IP data are switched at ATM speeds to the router, reassembled into IP packets, passed up to layer 3, a routing decision is made, then the IP packets are segmented into ATM cells again before being switched (maybe by the same switch as before), once again at ATM speeds.

See 2.2, "LAN Emulation Version 2.0" on page 59 for more details of LNNI.

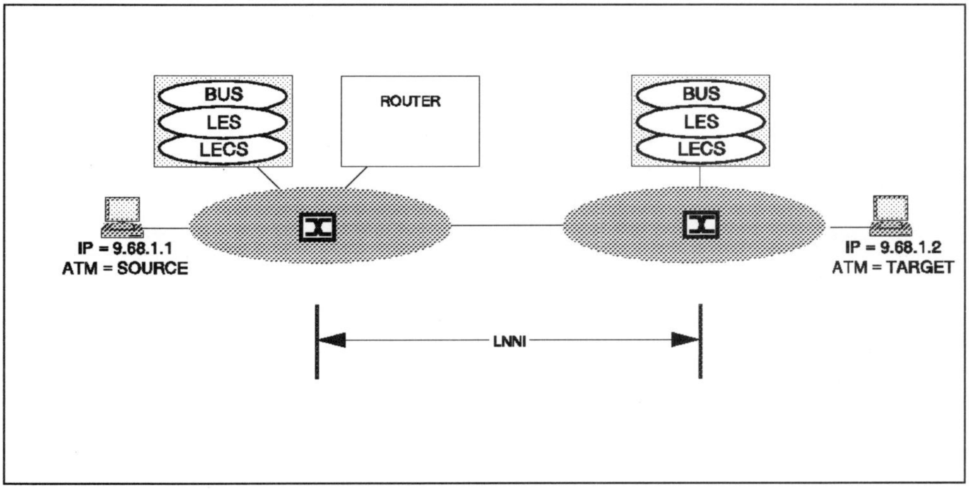

Figure 23. *Connecting Emulated LANs: Router Placement*

1.7.2.2 Router Networks

Legacy Based

Once routers are connected together, either a *dynamic routing* protocol or *static routes* are needed. This allows routers to communicate with each other and exchange information about the networks they know. Based on this information, a router decides which port to transmit a packet on towards the packet's ultimate destination. Well-known dynamic routing protocols are routing information protocol (RIP), open shortest path first (OSPF) and border gateway protocol (BGP).

Figure 24 shows a simple network of routers. If source 9.68.1.1 wants to send data to 9.72.1.1, then it can send the data to its default router 9.68.1.2, but how does the router know to send the data out on interface 9.70.1.1 if it wants to reach the 9.72.1 network? Routing tables built either by a dynamic routing protocol, or by definitions, allow the router to make such a decision. The route finally chosen is based on metrics such as hop-count or route-cost.

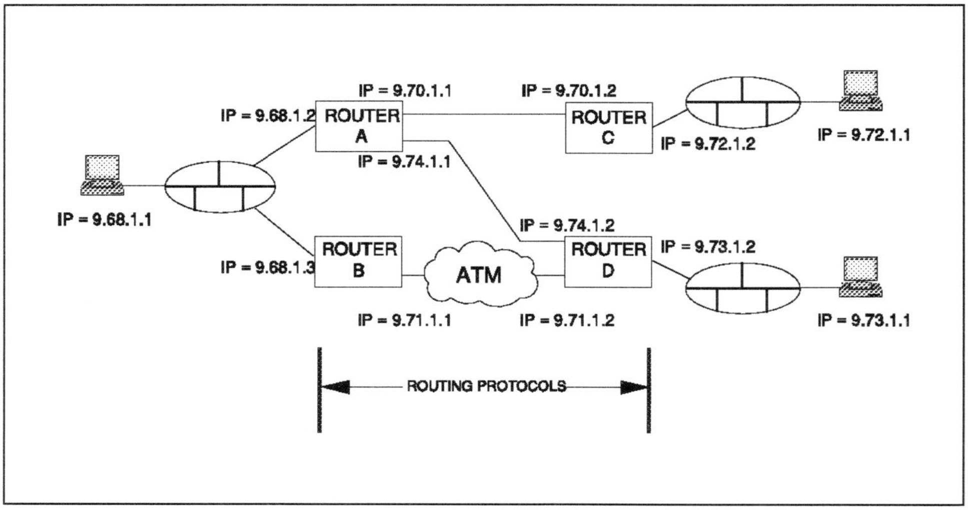

***Figure 24.** Routing Protocols*

ATM-Based

Private Network-to-Network Interface

In Figure 24, routers B and D are connected to an ATM network. This could be done by a LAN emulation layer, IP over ATM connection in the router, or by an external device. With all these solutions, the routing protocol in the router has no knowledge whatsoever about how to find an optimal route through the ATM network to a target router. There are no provisions for defining quality of service (QOS) or traffic management. To overcome these shortcomings, various solutions are being worked on. The ATM private network-to-network interface (PNNI) can be viewed as a routing protocol for ATM switches.

See Chapter 6, "PNNI Phase 1 and Integrated PNNI" on page 135 for more details of PNNI.

PNNI Augmented Routing

PNNI augmented routing (PAR) describes how ATM-attached routers will participate in PNNI. All routers in the network will run a dynamic IP routing protocol (for example RIP, OSPF, BGP or EGP). The routing information supplied by these protocols is used to forward IP packets. ATM switches and routers attached to the ATM network will run standard PNNI routing. PAR enables interaction between PNNI and the routing protocol in the router. PAR routers will announce to PNNI their status as routers and will use PNNI to locate other PAR routers through the ATM network. In addition, these routers will have detailed knowledge of the internals of the ATM network.

See 6.6, "PNNI Augmented Routing" on page 148 for more details of PAR.

Integrated PNNI

Integrated PNNI (I-PPNI) will replace the present routing tables with a single topology database containing ATM and layer 3 routing information. Routers and ATM switches will all have the same information about the network. A router will be able to make a routing decision based on the complete path to a target, no matter whether it goes through routers or ATM switches or any combination of both. The route chosen will take ATM QOS and traffic management parameters into consideration.

See 6.7, "Integrated PNNI (I-PNNI)" on page 150 for more details of I-PNNI.

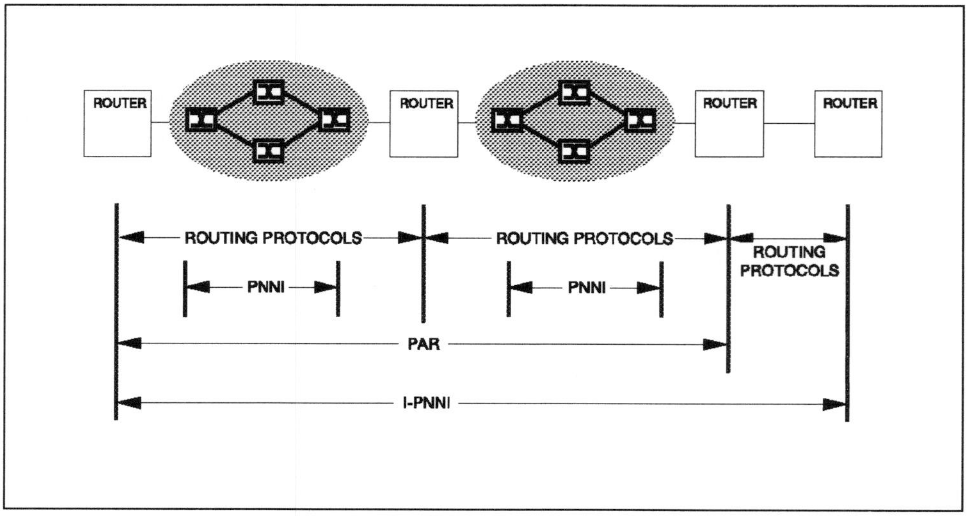

Figure 25. Scope of Routing Protocols

1.7.2.3 Switch Networks

ATM networks support two types of connections through them for data transfer, *permanent virtual circuits* (PVCs) and *switched virtual circuits* (SVCs). There is currently no defined standard for the setup of PVCs.

Static Paths

Permanent virtual circuits can be established either by an operator or by management interfaces when needed, or when a switch is activated. The endpoints and specifications of the PVC must be specified. PVCs generally stay active even when they are not being used. One exception is PVCs that are used for control and management between ATM switches. The setup of these connections usually occurs implicitly, for example, when a link between two ATM switches is activated.

Dynamic Paths

Switched virtual circuits make better use of network resources as they are usually only established when needed and taken down again when not needed. SVCs are usually initiated by applications in the network, and usually better meet the requirements of an application. To enable the dynamic setup of SVCs in a network of ATM switches, various interfaces and protocols are needed.

The interim local management interface (ILMI) is shown in Figure 26 on page 41. This is between the ATM end node and the ATM switch it is connected to. It uses the well-known VC VPI=0, VCI=16. The ILMI, among other things, allows an ATM end node, when activated, to register with the ATM switch it is connected to. The end node passes its end system identifier (ESI) to the ATM switch and, in return, gets the address prefix back from the switch. These two parts, along with the authority and format identifier (AFI), are used to build the end node's ATM address. In addition, through this registration process, the ATM switch now knows which port an ATM end node is connected to.

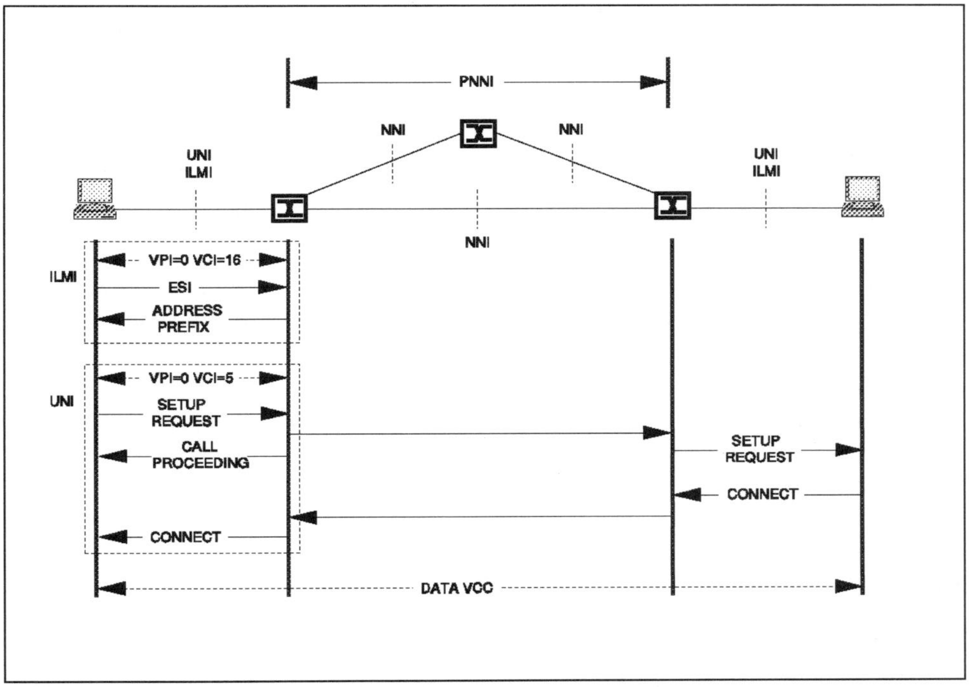

***Figure 26.** ATM Registration and Call-Setup*

The user-to-network interface (UNI) is defined between switches and ATM end nodes. The well-known default VC VPI=0, VCI=5 is used for UNI signalling. The network-to-network interface (NNI), which is an extension of UNI, is defined between ATM switches. ATM network topology is exchanged between ATM switches over the NNI using the private network-to-network interface (PNNI).

PNNI consists of two components:

PNNI Signalling Protocol
This is the signalling protocol used to relay ATM connection requests through the ATM network from the source UNI to the target UNI.

Virtual Circuit Routing Protocol

This is the protocol used to route a signalling request through the network from source to target. The path this request takes is the path that the virtual circuit will be established along, and along which data will then flow. The signalling request must be routed as there is no connection for it to flow along before the connection is set up.

1.7.2.4 Resource Reservation

APPN Networks

A connection-oriented network is one that predetermines a path through the entire transport before sending data. A connection network, due to transport reservation, state information and setup schemes, can allow optimal performance, manageability and stability. This is the reason why ATM is also a connection-oriented network.

Quality of service (QOS) guarantees can only be acheived when resources are reserved when the connection is set up. In addition, an application must be able to request the QOS it requires from the transport network.

Existing APPN applications already request a class of service (COS) from an APPN network. Native APPN ATM DLC support, described in 4.2.2, "Native ATM DLC" on page 87, will map an application's COS request to an ATM QOS. Applications will not need to be changed to take full benefit from ATM's QOS.

IP Networks

IP networks are connectionless. IP packets do not flow on predetermined paths. In fact, each IP packet may take a different path through a network. The Resource Reservation Protocol (RSVP) explained in 8.3, "Resource Reservation Protocol (RSVP)" on page 189 is being proposed as a method of reserving resources in an IP network. To reserve resources, packets will flow along the *expected* data path and attempt to reserve the resources needed by the application. In cases where the data path consists of a single VCC through an ATM network, RSVP will be capable of requesting a QOS end-to-end. RSVP goes further in that it would also attempt to reserve resources and bandwidth in non-ATM parts of a data path between two nodes.

To exploit RSVP fully, changes will need to be made to applications.

1.7.3 Multiprotocol Support

Chapter 3, "Multiprotocol over ATM (MPOA)" on page 65 describes perhaps the most wide reaching proposal for the use of ATM networks by existing layer 3 protocols. MPOA takes various solutions to parts of the problems of using IP in ATM networks, LAN emulation (LANE), Classical IP over ATM (IPOA), next hop resolution protocol (NHRP) and multicast address resolution server (MARS) and enhances and integrates them into a single solution. The MPOA specification, which is still under development, only addresses the needs of IP as the layer 3 protocol being supported.

IBM Multiprotocol Switched Services (MSS), discussed in 7.3, “IBM Multiprotocol Switched Services” on page 171, is IBM's implementation of MPOA with some functions added.

Chapter 2. Emulated and Virtual LANs

An ideal environment would only have ATM as the network transport layer and all applications would connect directly to this layer. Current network users have invested heavily in their physical networking infrastructure and applications which are built to use this infrastructure. The investment in existing networks means that a migration path must be offered to enable a gradual migration to ATM networks. In order to enable such *legacy* applications to use an ATM network without changes, it was neccessary to define a method whereby the applications can run unchanged, and the networking hardware can gradually be migrated to ATM. The ATM Forum LAN Emulation over ATM Specification V1.0 allows such a migration.

LAN emulation supports all layer 3 protocols transparently, while RFC 1577, or Classical IP and ARP over ATM (see 3 on page 229), is designed only to support IP as the layer 3 protocol.

In this chapter LAN Emulation Version 1.0 is explained followed by an overview of LAN emulation Version 2.0 draft specification. This is followed by a short description of virtual LANs (VLANs).

2.1 LAN Emulation Version 1.0

LAN emulation enables the implementation of emulated LANs over an ATM network. An emulated LAN provides communication of user data frames across of its users, similar to a physical LAN. One or more emulated LANs could run on the same ATM network. However, each of the emulated LANs is independent of the other and users cannot communicate directly across emulated boundaries, this is exactly the same as physical LANs. Inter-ELAN communication will only be possible through routers or bridges.

Each emulated LAN has only one type; either Ethernet/IEEE 802.3 or Token-Ring/IEEE 802.5. An emulated LAN is composed of several LAN emulation clients (LEC) and a single LAN emulation service entity (LE Service). The LE Service consists of an LE configuration server (LECS), an LE server (LES) and a broadcast and unknown server (BUS). The LE client resides in an ATM end station. It represents a set of users, identified by their MAC address. The LE Service may be part of an end station or a switch. It can be centralized or distributed over a number of stations.

Communication between LE clients, and between LE clients and the LE Service, is performed over ATM virtual channel connections (VCCs). Each LE client must communicate with the LE Service over control and data VCCs. Emulated LANs operate in any of the possible environments:

- Switched virtual circuit (SVC)
- Permanent virtual circuit (PVC)
- Mixed SVC and PVC

In a PVC-only LAN, there are no call setup and close down procedures. Instead of that, layer management is used to set up and clear the connections. In this PVC environment, the layer management is responsible for both setting up and clearing connections and has the responsibility that the emulated LAN works correctly.

2.1.1 LAN Emulation Protocol Stack

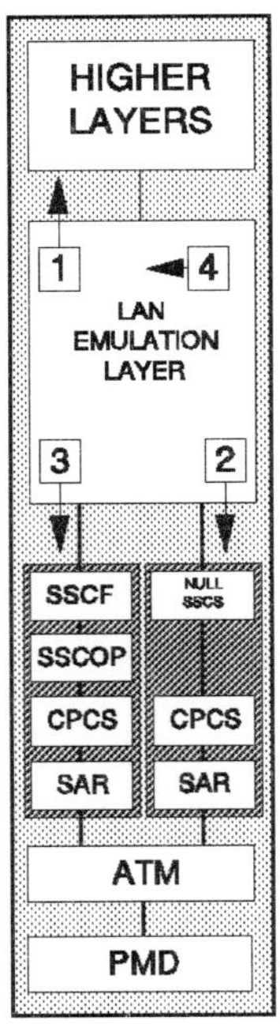

Legend:

- Higher Layers (802.2, NetBIOS, IPX,...)
- Control Plane (SAAL)
  - SSCF Service Specific Coordination Function
  - SSCOP Service Specific Connection Oriented Protocol
  - CPCS Common Part Convergence Sublayer
  - SAR Segmentation and reassembly Sublayer
- User Plane (AAL5)
  - SSCS Service Specific Convergence Sublayer
  - CPCS Common Part Convergence Sublayer
  - SAR Segmentation and reassembly Sublayer
- ATM ATM Layer
- PMD Physical Medium Dependent Layer

Figure 27. LANE Emulation Layers

LAN emulation over ATM operates within the data link layer of the OSI reference model. In Figure 27 the position of the LAN emulation layer is shown.

The following list explains the flows depicted by the arrows in the figure:

1. LE to Higher Layer Services

   These services apply only to the LAN emulation client. The higher layer could be logical link control, or equivalent, or a bridging relay function. The services provide the capability to exchange user data frames over the LAN emulation service. Service definitions are compatible with ISO 10039 service architecture and ISO 10038 MAC bridging standard.

2. LAN Emulation to AAL Services

   These services apply to the LAN emulation clients and the LAN emulation service. These services provide the capabilities to transfer frames between peer LAN emulation layers. This specification assumes a null Service Specific Convergence Sublayer (SSCS), that is, the SSCS provides for the mapping of the equivalent primitives of the AAL and the Common Part Convergence Sublayer (CPCS). See also Figure 68 on page 193 in Appendix A, "Protocol Stack Reference" on page 193. The common part of AAL5 makes use of the services provided by the underlaying ATM layer.

   A LAN emulation entity includes the following AAL service interfaces, each identified by a distinct SAP-ID. Each LAN emulation client includes the following SAPs:

   a. One or two control SAPs that handle initialization, registration and address resolution

   b. Two or more data forwarding SAPs

   c. Zero or one control SAP that handle configuration

3. Connection Management Services

   These services apply to the LAN emulation clients and the LAN emulation service.

   The conceptual model assumed by the LAN emulation layer is shown in Figure 27 on page 46. The Connection Management may use either PVCs or SVCs and provides the following primitives:

   - Setup

     This service provides initial call establishment. It receives an ATM address and establishes a virtual connection identified by a SAP-ID.

   - Release

This service is used to request the network to clear an end-to-end connection identified by a SAP-ID.

- Add Party

  This service provides the capability to add a party to an existing connection.

  This service is used to drop or clear a party from an existing point-to-multipoint connection.

4. LAN Emulation to Layer Management Services

   These services enable initialization and control of the LAN emulation entities. These services differ between LAN emulation clients and the LAN emulation service.

2.1.2 LAN Emulation Components

An emulated LAN consists of several components. Clients, for example, ATM workstations and ATM bridges, each having at least one LE client entity. The LE Service consists of several components, including the LE server, the broadcast and unknown server, and the LAN emulation configuration server.

2.1.2.1 LAN Emulation Client (LEC)

The LAN emulation client is the entity in the end systems that performs data forwarding, address resolution, and other control functions. This provides MAC level emulated Ethernet/IEEE 802.3 or IEEE 802.5 service interface to higher-level software and implements the LUNI interface when communicating with other entities within the emulated LAN.

2.1.2.2 LE Server (LES)

The LE server implements the control coordination for the emulated LAN. The LE server provides a facility for registering and resolving MAC addresses and/or descriptors to ATM addresses. Clients may register the LAN destinations they represent with the LE server. A client will also query the LE server when the client wishes to resolve a MAC address and/or route descriptor to an ATM address. The LE server will either respond directly to the client or forward the query to other clients so they may respond.

2.1.2.3 Broadcast and Unknown Server (BUS)

The broadcast and unknown server (BUS) handles data sent by an LE client to the broadcast MAC address ('FFFFFFFFFFFF'). All multicast and initial unicast frames that are sent by a LAN emulation client before the data direct VC target ATM address has been resolved (before a data direct VCC has been established) are handled by the broadcast and unknown server.

The multicast function provided in the BUS may be implemented by an underlying ATM multicast service. The BUS multicast function must be consistent with ITU-T Recommendation X.6 Multicast Service Definition.

A LAN emulation client sends data frames to the BUS, which serializes the frames and retransmits them to a group of attached LAN emulation clients. Serialization is required to prevent AAL-5 frames from different sources from being interleaved.

In an SVC environment, the BUS needs to participate in the LE address resolution protocol (LE_ARP) to enable a LAN emulation client to locate the BUS.

The BUS must always exist in the emulated LAN and all LAN emulation clients must join its distribution group.

2.1.2.4 LE Configuration Server (LECS)

An ATM network can consist of several emulated LANs. The LE configuration server assigns LE clients to an emulated LAN based on its configuration database, its own policies and the information it receives from the respective LE clients. It assigns any client that requests configuration information to a particular emulated LAN service by giving the client the LES's ATM address. This method supports the ability to assign a client to an emulated LAN based on either the physical location (ATM address) or the identity of a LAN destination that it is representing.

It is optional for the LAN emulation client to obtain information from the LECS using the configuration protocol. The LECS allows the LAN emulation client to be automatically configured.

2.1.2.5 LAN Emulation Connections

VCCs are used for the connections of LAN emulation clients and other entities of LAN emulation such as the LECS, LES and BUS. For each different connection a separate VCC exists. In Figure 27 on page 46, the connections are shown with the stronger lines designating control connections.

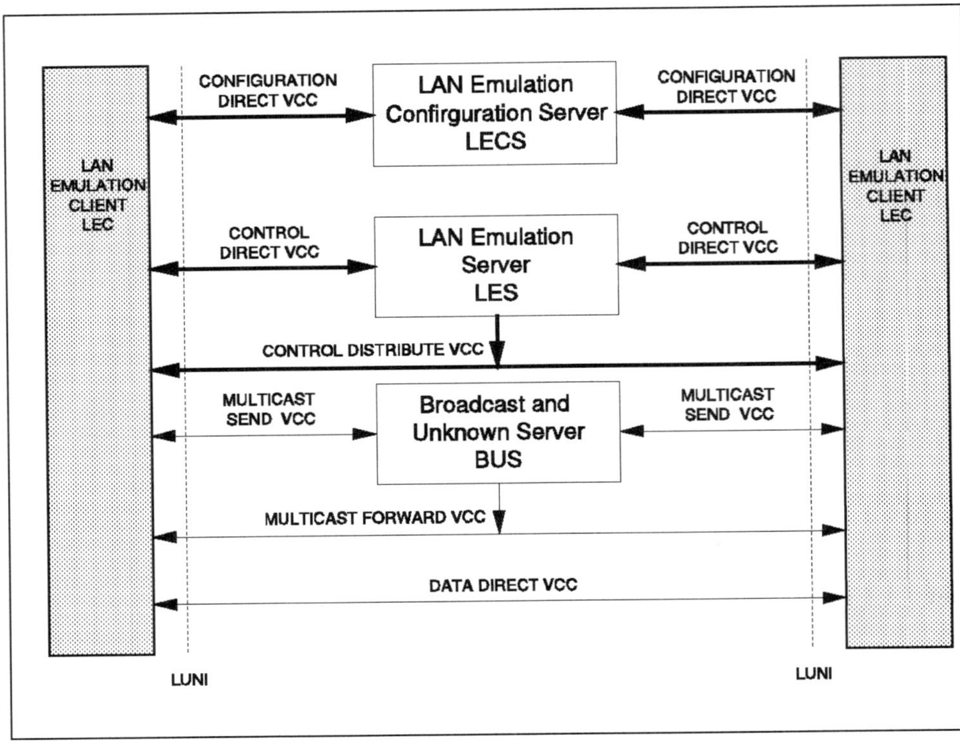

Figure 28. LANE Components

Control Connections

Several control VCCs exist. They link the LEC to the LECS, or link the LEC to the LES, and carry LE_ARP traffic and control frames. A control VCC never carries data frames. Building control VCCs is a part of the LAN emulation client initialization process.

Three different control VCCs exist:

Configuration Direct VCC

The configuration direct VCC is a bidirectional VCC that may be set up by the LAN emulation client, or other entity, as part of the LECS connect phase. It is used for obtaining configuration information, including the address of the LES. The entity may maintain this VCC while participating in the emulated LAN. It may continue to keep it open for further queries to the LECS while participating in the emulated LAN. The configuration direct VCC may be used to inquire about an LE client other than the one to which the configuration direct VCC is attached. This connection is signaled using B-LLI to indicate it carries LE Control packet formats.

Control Direct VCC

The LAN emulation client sets up a bidirectional point-to-point VCC to the LES for sending control traffic. This is set up by the LAN emulation client as part of the initialization phase. The LES has the option to use the return path for sending control data to the LAN emulation client. The LAN emulation client must thus accept control data from this VCC.

The LEC and LES must maintain this VCC as long as they are members of the emulated LAN.

Control Distribute VCC

The LES may optionally set up a unidirectional point-to-point or point-to-multipoint VCC to the LAN emulation client for distributing control traffic. This VCC may be set up by the LES as part of the initialization phase. If this VCC is set up, the LAN emulation client must accept the control distribute VCC.

The LEC and LES must maintain this VCC as long as they are members of the emulated LAN.

Data Connections

Data connections are used to connect LECs to LECs, and LECs to the broadcast and unknown server. The VCCs carry IEEE 802.3 or IEEE 802.5 data frames as well as flush messages. A flush message is the only time that a data connection will have control traffic. A flush message is generated by the flush protocol in order to ensure that the frames will remain in sequence when two data paths exist. This possibility exists when data is sent both via the broadcast and unknown server and via a direct VCC.

The following VCCs are defined:

Data Direct VCC

A data direct VCC is a bidirectional point-to-point VCC between LECs that wants to exchange unicast data traffic.

When a LAN emulation client has a packet to send and the ATM address for the destination MAC address is unknown, the LEC generates an LE_ARP request to resolve the ATM address for the destination. Once the LEC receives a reply to the LE_ARP, it sets up a point-to-point VCC, if not already established, over which to send all subsequent data to that destination.

The LEC that issues an LE_ARP request and receives an LE_ARP response is responsible for initiating the signalling to establish a bidirectional data direct VCC with the LEC sought in the LE_ARP request.

Multicast Send VCC

This VCC is used for sending multicast data to the BUS and for sending initial unicast data. The BUS may use the return path on this VCC to send data back to the LEC. This requires that the LEC accept traffic on this VCC.

A LAN emulation client sets up a bidirectional point-to-point multicast send VCC to the BUS. This VCC has the same setup as the data direct VCC. The LEC first sends an LE_ARP request and, when it receives the LE_ARP response, initiates signalling to establish the multicast send VCC to the BUS.

The LAN emulation client must maintain this VCC as long as it is a part of the emulated LAN.

Multicast Forward VCC

The multicast forward VCC is initiated by the broadcast and unknown server. This is done after the LAN emulation client has set up the multicast send VCC. The multicast forward VCC is used for distributing data from the broadcast and unknown server. It can be either a point-to-multipoint VCC or a unidirectional point-to-point VCC. The LEC emulation client must accept the multicast forward VCC regardless of type. A multicast forward VCC from the broadcast and unknown server must be established before a LAN emulation client can participate in the emulated LAN.

The LAN emulation client must attempt to maintain this VCC as long as it is a member of the emulated LAN.

The broadcast and unknown server may forward frames to a LAN emulation client on either the multicast send VCC or the multicast forward VCC. A LAN emulation client will not receive duplicate frames forwarded from the broadcast and unknown Server on both the VCCs, but must be able to accept frames on either VCC.

2.1.3 LAN Emulation User-to-Network Interface

LE clients and the LE Service must interact in a well defined manner. This is accomplished using PDUs and well defined protocols. Four steps can be distinguished:

Initialization

- Obtaining the ATM address(es) of the LE Services that are available in an ATM network
- Joining or leaving a particular LAN specified by the ATM address of the LE Service

- Declaring whether this LE client wants to receive address resolution request for all the frames with unregistered destinations

Registration
Informing the LE Service of the following:

- The list of individual MAC addresses that the LEC client represents
- The list of source route descriptors (for example, segment/bridge pairs) that the LE client represents for source route bridging

Address resolution
Obtaining the ATM address representing the LE client with a particular MAC address (unicast, broadcast or segment/bridge pair).

Data transfer:
Moving the data from the source to the destination by:

- Encapsulation of the LE-SDU (service data unit) in an AAL-5 frame and transmission by the LE client
- Forwarding of the AAL-5 frame by the LE Service (if applicable)
- Reception and header removal of the AAL-5 frame by the LE client

2.1.4 LAN Emulation Functions

LAN emulation service is divided into seven functions. Some of the functions can be divided into several subfunctions. This chapter describes the functions and subfunctions as defined by the ATM Forum.

2.1.4.1 Initialization

Several steps are taken before the Initialization function is completed:

Initial State
In the initial state, there are parameters (for example, addresses, emulated LAN, maximum frame size, etc.) that are known to the LE server and the LE clients about themselves before they participate in the configuration and join phase functions.

LAN Emulation Configuration Server Connect Phase
In the LECS connect phase, the LE client establishes a configuration direct VCC to the LE configuration server.

Join Phase
In the join phase of ATM LAN emulation initialization, the LAN emulation client establishes its control connections to the LAN emulation server. The join phase can have two outcomes: success or failure.

Once the join phase has successfully completed, the LAN emulation client has been assigned a unique LAN emulation client identifier (LECID). It now knows the emulated LAN′s maximum frame size and its LAN type. It also has established the control VCC(s) with the LAN emulation server.

Initial Registration

After joining, a LAN emulation client may register any number of MAC addresses and/or route descriptors. This is in addition to the single MAC address that can be registered as part of the join phase. Initial registration allows a LAN emulation client to verify the uniqueness of its local addresses before completing initialization and becoming operational.

Connecting to the Broadcast and Unknown Server

In order to establish a connection to the broadcast and unknown server, the LAN emulation client LE_ARPs for the broadcast MAC address and proceeds to set up the connection. The broadcast and unknown server then establishes the multicast forward VCC to the LAN emulation client.

Initialization Phases, Recovery and Termination

2.1.4.2 Registration

The address registration function is the mechanism by which clients provide address information to the LAN emulation server. An intelligent LAN emulation server may respond to address resolution requests if LAN emulation clients register their LAN destinations, defined as MAC addresses or, for source routing IEEE 802.5 LANs only, route descriptors, with the LAN emulation server. The LAN destinations may also be unregistered as the state of the client changes. A client must either register all LAN destinations for which it is responsible or join as a proxy.

2.1.4.3 Address Resolution

Address resolution is the procedure by which a client associates a LAN destination with the ATM address of another client or the broadcast and unknown server. Address resolution allows clients to set up data direct VCCs to carry frames.

When a LAN emulation client is presented with a frame for transmission whose LAN destination is unknown to that client, it must issue a LAN emulation address resolution protocol (LE_ARP) request frame to the LAN emulation server over its control point-to-point VCC.

The LAN emulation server may either:

1. Forward this LE_ARP to the appropriate client(s) using the control distribute VCC or one or more control direct VCCs. Different LAN emulation server implementations may use different distribution algorithms. If a client responds to a forwarded

LE_ARP request with a LE_ARP reply, that reply is also sent and forwarded over the control VCCs to the original requestor.

2. Or instead of forwarding the LE_ARP, the LAN emulation server may issue an LE_ARP reply on behalf of a client that has registered the requested LAN destination with the LAN emulation server.

A LAN emulation client must respond to an LE_ARP that it receives, asking for a LAN destination it has registered with the LAN emulation server or for which it is a proxy.

Each LAN emulation client maintains a cache of LE_ARP replies and uses a two-period time-out mechanism to age entries in this cache. The aging time period is used for all entries learned from LE_ARP responses who's remote address flag was zero. That is, responses for registered LAN destinations are always timed out with the aging time. For aging entries learned from LE_ARP replies with the remote address FLAGS bit set to 1 and for entries learned from observing source addresses on data VCCs, which time-out to use is determined by the state of the LAN emulation client's topology change flag. When this flag is set, such entries are aged using the aging time parameter. The state of this flag may be altered either by local management or by reception of the LE_TOPOLOGY_REQUEST messages.

2.1.4.4 Connection Management

In switched virtual connection environments, the LAN emulation entities set up connections between each other using UNI signalling. The connections use best-effort quality of service as the minimum level.

Call establishment

When a call is being set up, the destination must not send its CONNECT message until it is ready to receive frames on the new VCC. The originator should expect that it can transmit frames after it has received the CONNECT message from the destination.

The CONNECT_ACK message is received by the destination and can be generated by its local switch. This message can reach the destination before the CONNECT message reaches the originator. The originator can only start to initialize its VCC after it receives the CONNECT message from the destination. Therefore, there is no guarantee for the destination that its initial data will be received by the originator until it receives some end-to-end indication from the originator.

The originator must send a READY_IND message as soon as it is ready to receive frames on the newly established VCC. At that point, the originator considers call establishment to be complete. The originator may also send data as soon as it is ready to receive frames on the newly established VCC. Data may be sent before or after sending the READY_IND.

It is possible that the READY_IND message can get lost. To recover it, the destination is responsible for timing the arrival of the READY_IND message. If the timer expires, the destination sends data or a READY_QUERY message on the VCC. Either party shall always respond to receipt of a READY_QUERY message on an active VCC by transmitting a READY_IND message.

Tear down and timeout of VCCs

If a control direct VCC or control distribute VCC is ever released, a LAN emulation client must always return to the LAN emulation configuration server connect phase of initialization. If the broadcast and unknown server VCC is lost while the LAN emulation client is participating in a emulated LAN, the LAN emulation client may return to the broadcast and unknown server connect phase or go to the termination phase and restart.

2.1.4.5 Data Transfer

There are two different connections used for data transfer:

1. Data direct VCCs between individual LAN emulation clients
2. Multicast send and multicast forward VCCs that connect clients to the broadcast and unknown server

Unicast Frames

When a LAN emulation client has established, via the address resolution mechanism, that a certain LAN destination corresponds to a certain ATM address, and when that client knows is has a data direct VCC to that ATM address, then a frame addressed to that LAN destination must be forwarded via that data direct VCC.

If a LAN emulation client does not know which data direct VCC to use for a given unicast LAN destination, or if that data direct VCC has not yet been established, it may elect to transmit the frame over the multicast send VCC to the broadcast and unknown server. The broadcast and unknown server, in turn, forwards the frame to at least the client for which it is destined. If the LAN destination is unregistered, then the frame must be forwarded to at least all proxy clients and may be forwarded to all clients.

On an emulated LAN, the case can arise where a frame can only reach its destination through an IEEE 802.1D transparent bridge, and that bridge does not know the whereabouts of that destination. The only way such a frame can be assured of reaching its destination is for the frame to be transmitted to all of the IEEE 802.1D transparent bridges via the broadcast and unknown server so that they, in turn, can flood that frame to all of their other bridge ports, or at least the ones enabled by the spanning tree protocol. A LAN emulation client that chooses not to forward frames to the broadcast and unknown server, therefore, may not be able to reach destinations via transparent bridges, or perhaps other proxy agents.

Multicast Frames

LAN emulation clients may wish to send frames to a multicast MAC address, and/or they may wish to receive frames addressed to a given multicast MAC address. In order to send frames to a multicast MAC address, a LAN emulation client must send the frames to the broadcast and unknown server. The address resolution mechanism is used during the initialization process to provide the ATM address of the broadcast and unknown server for multicast and broadcast traffic, and connection management will provide a point-to-point multicast send VCC over which to send such frames.

All that is required in order for the LAN emulation client to receive frames addressed to a given multicast MAC address is for the LAN emulation client to connect to the broadcast and unknown server, after which the broadcast and unknown server will try to set up a return path for all broadcast and multicast traffic. When a client connects to the broadcast and unknown server, the broadcast and unknown server will try to establish a multicast forward VCC to that client. It is expected that multicast forward VCCs will be unidirectional point-to-multipoint VCCs, but they may be implemented as point-to-point VCCs. This decision is left to the LAN emulation service, not to the client.

A LAN emulation client will receive all flooded unicast frames and all broadcast and multicast frames over either its multicast send VCC or its multicast forward VCC. Which VCC the broadcast and unknown server uses to forward frames to the LAN emulation client is at the discretion of the broadcast and unknown server. A LAN emulation client will not, however, receive duplicate frames.

The LAN emulation header of any data frame sent from a client to the broadcast and unknown server must either contain the value 0 or the unique LECID value assigned to that client. The broadcast and unknown server is required to preserv the LAN emulation header of a relayed frame. Thus, a client can identify and filter frames that it sent by comparing the LECID field to its own LECID value. A transparent bridge LAN emulation client cannot reliably use the source MAC address to identify its own broadcast and unknown server traffic.

Token-ring functional addresses are treated just as any other multicast MAC address.

2.1.4.6 Frame Ordering

There may be two paths for unicast frames between a sending LAN emulation client and a receiving client: one via the broadcast and unknown server and one via a data direct VCC between them. For a given LAN destination, a sending client is expected to use only one path at a time, but the choice of paths may change over time. Switching between those paths introduces the possibility that frames may be delivered to the receiving client out of order. Out-of-order delivery between two LAN endpoints is uncharacteristic of LANs and undesirable in an ATM emulated LAN. The flush protocol is provided to ensure the correct order of delivery of unicast data frames.

2.1.4.7 Source Route Considerations

Source route bridging is the predominant bridging technology used within IEEE 802.5 token-ring networks. The use of source routing does not preclude transparent bridging in these networks. A token-ring end station will typically use a combination of source-routed and nonsource-routed frames. This allows a LAN emulation client to operate with both source routing and transparent bridging.

In addition to the Destination Address (DA) field and Source Address (SA) field, a source-routed frame contains a Routing Information (RI) field. The RI field contains a control field and a list of route descriptors (RD) that indicate the frame′s path through the network. Therefore, the information in the RO field determines which SR bridges will forward the frame. The LAN emulation client determines if the frame is to be forwarded by an SR bridge or if the LAN destination is a station on the emulated LAN.

The LAN emulation client determines if the frame is to be forwarded by an SR bridge, or if the LAN destination is a station on the local emulated LAN by examining the RI field. If the LAN destination is accessible through an SR bridge, the LAN destination is the Next Route Descriptor (Next_RD), otherwise, the LAN destination is the frame′s destination address.

Frames with specifically routed source routing information (an SRF frame) and unicast destination MAC address are sent down data direct VCCs following the usual LE_ARP and VCC setup process. Other source-routing frames are sent through the broadcast and unknown server.

2.2 LAN Emulation Version 2.0

As a follow up to LANE Version 1.0, the ATM Forum is working on LAN Emulation over ATM Version 2. The LANE V2 specifications are split into two separate work items.

The LANE V1 specification details the interface between LAN emulation client and LAN emulation server. Work on the LUNI specification is primarily an extension to LANE V1.

A new LNNI specification defines the interface between LAN emulation server entities.

LUNI Specification

In the connect phase between the LES and LEC, the following order was used to determine the ATM address of the LECS:

1. Get the LAN emulation configuration server address via the ILMI
2. Use the well-known LECS address
3. Use the LECS PVC

In the LUNI V2 specification, the LEC will try to contact the LECS using a preconfigured LECS address, if that is available. It is not mandatory to configure the LEC with this address. If it is configured, the LEC must attempt to establish the configuration direct VCC to that address. If this fails or when the address is not defined, the three methods used in LUNI V1 must be followed in the predefined order.

LNNI Specification

LANE V1 only defines communication between the LAN emulation client and the LAN emulation service entities. The LNNI specification deals with the communication between LAN emulation server entities. This communication covers LES-to-LES, BUS-to-BUS, LECS-to-LES and LECS-to-BUS connections. LES-to-BUS communication is not a part of the model, since the LES and the BUS always operate as a pair.

2.3 Virtual LANs

A virtual LAN (VLAN) is a logical grouping of users and servers independent of physical location. It can also be described as a group of users on a single broadcast domain; that is, broadcast traffic must be limited to members of a VLAN.

The concept of a *virtual workgroup* is supported by VLANs. A virtual workgroup is a group of people who need to communicate with each other, but who are in different locations. They need the same level of communication as if they were on the same LAN, but may be spread across different LANs, switches and WANs. Users can belong to multiple VLANs at the same time.

VLANs give network administrators the ability to dynamically configure virtual workgroups, adding, deleting and moving users quickly without making any physical changes to the networks. In addition, users can still belong to the same VLAN after they have moved locations.

Network management will allow a network administrator to set up multiple emulated LANs (ELANs) on a single ATM switch; each ELAN is effectively a VLAN in this case. The ATM switch port or layer 3 protocol address can also be used to differentiate between members of VLANs.

The IEEE 802 Executive Committee recently approved the development of a new virtual LAN (VLAN) interoperability standard to be known as 802.1q. The standard will go beyond earlier approaches to VLAN interoperability, including 802.10, in that it covers not only frame formats, but also rules for mapping packets to VLANs, packet forwarding, loop detection protocols, quality-of-service parameters, management architecture and MIBs. The result will be a flexible, comprehensive solution for interoperable VLANs, making them easier to define and to deliver better performance. Wide-scale adoption of distributed multivendor VLANs will be limited until the standards are in place.

The IEEE 802.1q standard will propose adding a four-byte explicit tag to each frame. The four-byte header will be more easily implemented in hardware than the up to 26-byte 802.10 header. Work also must be done on how to share VLAN information between switches. It is expected that the 802.1q specification will take 15 to 18 months to complete.

The following is a brief description of approaches for building VLANs.

2.3.1 VLAN Frame Tagging

After a switch has defined the appropriate VLAN for a sending device, it needs a method for identifying to other switches which VLAN the transmitted packets belong to. This issue, referred to as tagging, can be tackled by a number of methods:

- Explicit tagging
- Implicit tagging

Either a frame is explicitly tagged with the identity of the VLAN to which it belongs, or the frame contains an implicit tag, which requires that the frame be examined by each receiving port, and pattern matching/filtering applied to determine its VLAN membership. Both methods are required of any interoperable VLAN standard, that is, all VLAN switches on a given VLAN must agree to use the same tagging mechanism.

2.3.1.1 Explicitly-Tagged Frames

An explicitly-tagged, encapsulated frame may contain information including:

- Source and destination MAC address
- A tag value that identifies which VLAN the encapsulated frame belongs to
- An indication of the frame type format, for example, 802.3/Ethernet or 802.5 format

There are several possible methods one could use to create the encapsulated frame:

- Two-layer bridging
- MAC address encoding of exit port

Two-Layer Bridging

Using the two-layer bridging approach, a MAC address is assigned to each physical port on each VLAN switch that is used for sending or receiving explicitly-tagged frames. Whenever a VLAN switch encapsulates a frame, it uses the port's MAC address as the source MAC address. For each user-level destination MAC address on each VLAN, the VLAN switch maintains the MAC address of the bridge that sourced the last frame received from that user-level MAC address in its bridging table. Thus, each VLAN switch keeps track of the VLAN switch that each user-level MAC address is *behind*. This is similar to the way that transparent bridges keep track of the port on which a given MAC address was last received. When the bridge MAC address is not known, an all 802.1 bridges multicast address is used for which all VLAN bridges must be able to receive.

One advantage of this scheme is that it improves the scaling ability of the bridged network; a VLAN switch needs to know the MAC addresses of its local users, the MAC addresses of the users to which they are conversing, and the MAC addresses of all of the other bridges. It does not need to know the MAC addresses of all of the users in the network.

MAC Address Encoding of Exit Port

Using the two-layer bridging model just described, suppose that a VLAN switch assigns multiple MAC addresses to be used by each port sending or receiving explicitly-tagged frames.

That is, traffic bridged from port *n* to explicitly-tagged port *m* uses source MAC address *MAC(m,n)*. To the other VLAN switches, this VLAN switch now appears to consist of a number of different VLAN switches, one for each of this VLAN switch's other ports.

This scheme has the advantage that, when receiving frames, the VLAN switch can behave like multiple transparent bridges.

IEEE 802.10 VLANs

The IEEE has defined the *802.10 Interoperable LAN/MAN Security (SILS) Standard*. Conceived as a security protocol to protect IEEE 802 LANs and MANs (metropolitan area networks), it is an OSI layer 2 protocol that incorporates authentication and encryption techniques. The 802.10 standard defines a *secure data exchange* (SDE) protocol data unit (PDU) (see Figure 29 on page 63 for details), which allows for the secure interchange of data at the layer 2 level. This is a MAC layer frame with an 802.10 header inserted between the MAC header and the data. The 802.10 header is divided into different parts, each to implement a different function, with support for each portion of this header being optional.

Certain ATM and LAN manufacturers now use the *Security Association Identifier* (SAID) field of the 802.10 header as the VLAN ID. In such VLANs, packets have the 802.10 header added as they are forwarded onto a backbone. Other switches on the backbone filter out packets that do not have the correct VLAN ID. Switches in the same VLAN remove the 802.10 header before passing the packet on to stations on the VLAN.

Although IEEE 802.10 is a standard, it is a *security* standard. Using fields in the 802.10 header for VLAN tagging is proprietary.

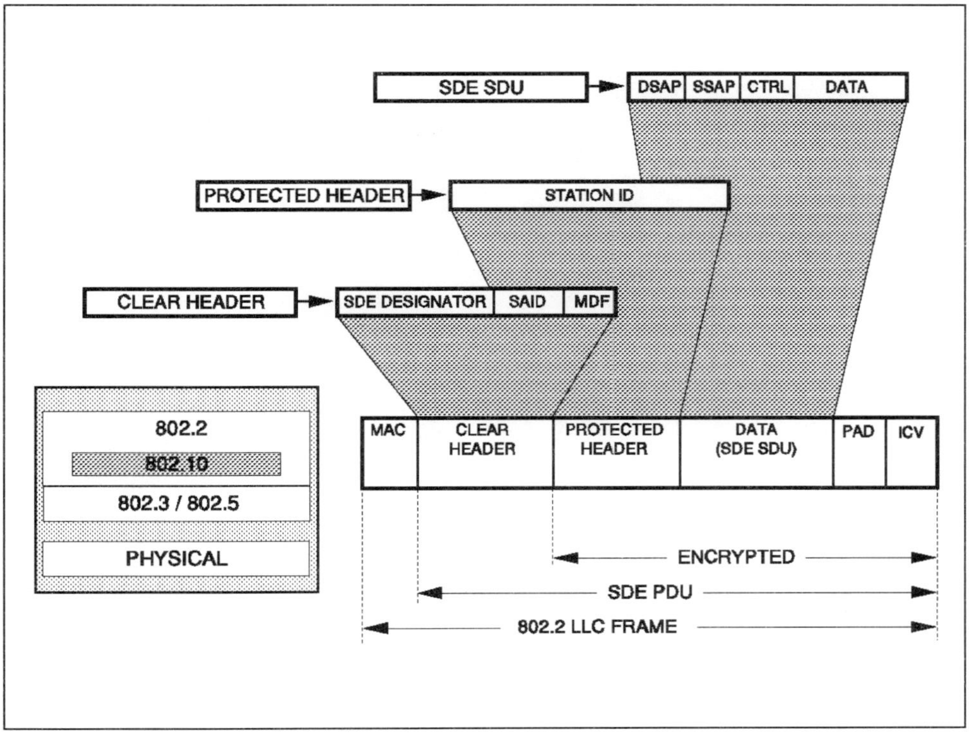

Figure 29. Structure of the SDE PDU

2.3.1.2 Implicitly-Tagged Frames

Protocol VLANs

One method of implicit tagging is to look into the layer 3 header inside each frame. In the case of IP, examination of the IP address would allow all members of an IP subnet to belong to the same VLAN.

VLAN Signalling

In the VLAN signalling approach, an ATM switch must maintain lists of which physical ports are associated with which VLAN. While this preserves the format of the original packet, when a VLAN spans more than one ATM switch, there must be some sort of protocol between the switches to exchange MAC to VLAN address mapping information. Another drawback is that unique MAC addresses are required across ATM switches.

Chapter 3. Multiprotocol over ATM (MPOA)

The specification of MPOA is being developed by the multiprotocol subworking group of the ATM Forum. At the time of writing, the status of the Specification Development for MPOA V1.0 is work in progress; that is, the work is at an early stage. A straw ballot is targeted for December 1996. Hence, the earliest date for final approved specification is April 1997.

This chapter is based on the latest specification of MPOA, (see 11 on page 230).

The objectives of MPOA are to:

- Provide end-to-end layer 3 internetworking connectivity across an ATM network. This is for hosts that are attached either:
  - Directly to the ATM network
  - Indirectly to the ATM network on a legacy LAN
- Support the distribution of the internetwork layer (for example, an IP subnet) across legacy and ATM-attached devices.
  - Removes the port to layer 3 network restriction of routers to enable the building of protocol-based virtual LANs (VLANs).
- Ensure interoperability among the distributed routing components while allowing flexibility in implementations.
- Address how internetwork layer protocols use the services of an ATM network.

Although the name is multiprotocol over ATM, the actual work being done at the moment in the MPOA subworking group is entirely focussed on IP. Token lip service is paid to other protocols such as IPX, AppleTalk and APPN. The official reasoning behind this policy is that it is easier to focus on one protocol and then add support for others later on.

3.1.1 Benefits of MPOA

MPOA represents the transition from LAN emulation to direct exploitation of ATM by the internetwork layer protocols. The advantages are:

- Protocols would see ATM as more than just another link.
  - Hence we are able to exploit the facilities of ATM.
- Eliminates the need for the overhead of the legacy LAN frame structure.

See Chapter 2, "Emulated and Virtual LANs" on page 45 for more information on LAN emulation.

The MPOA solution has the following benefits over both Classical IP (RFC 1577) and LAN emulation solutions:

- Lower latency by allowing direct connectivity between end systems that can cut across subnet boundaries (see 5.2, "IP Address Resolution in ATM Networks" on page 114). This is achieved by minimizing the need for multiple hops through ATM routers for communication between end systems on different virtual LANs.
- Higher aggregate layer 3 forwarding capacity by distributing processing functions to the edge of the network.
- Allows mapping of specific flows to specific QOS characteristics. For example, a layer 3 switch could interpret a layer 3 packet header in RSVP 13 on page 230 and use it to set up an ATM connection with the appropriate QOS.
- Allows a layer 3 subnet to be distributed across a physical network.

3.1.2 Technology Used by MPOA

LAN Emulation

- Intra-IASG traffic flowing between legacy media or LANE ATM devices is sent using the services of LANE.
- Multiple ELANs per IASG are permitted, as are multiple IASGs per ELAN.
- Mapping between IASGs and emulated LANs is outside the scope of MPOA.

Note: The configurations shown in Figure 31 on page 72 and Figure 32 on page 73 both show communication between ATM-attached hosts and LEC-connected hosts. In these cases, as shown, an LEC is required in ATM-attached hosts, or a DFFG may provide the functionality to communicate with LECs if no LEC is installed in the AHFG. Communication between ATM-attached hosts does not require use of the LEC.

Next Hop Resolution Protocol

- Being developed by the IETF to support the establishment of direct connections that cut across subnet boundaries
- Inter-IASG address resolution based on NHRP

Multicast Address Resolution Server / Multicast Connection Server

- Being developed by IETF to support IP multicast traffic within a Classical IP subnet on ATM.
- Internetwork layer multicast will use a generalized MARS/MCS function.

3.2 MPOA Logical Components

The MPOA solution consists of a number of logical components and information flows between those components. The logical components are of two kinds:

MPOA Server
MPOA servers maintain *complete* knowledge of the MAC and internetworking layer topologies for the IASGs they serve. To accomplish this, they exchange information among themselves and with MPOA clients.

MPOA Client
MPOA clients maintain local caches of mappings (from packet prefix to ATM information). These caches are populated by requesting the information from the appropriate MPOA server on an as-needed basis.

The layer 3 addresses associated with an MPOA client would represent either the layer 3 address of the client itself, or the layer 3 addresses reachable through the client. (The client has an edge device or router.)

An MPOA client will connect to its MPOA server to register the client's ATM address and the layer 3 addresses reachable via the client.

3.3 MPOA Functional Components

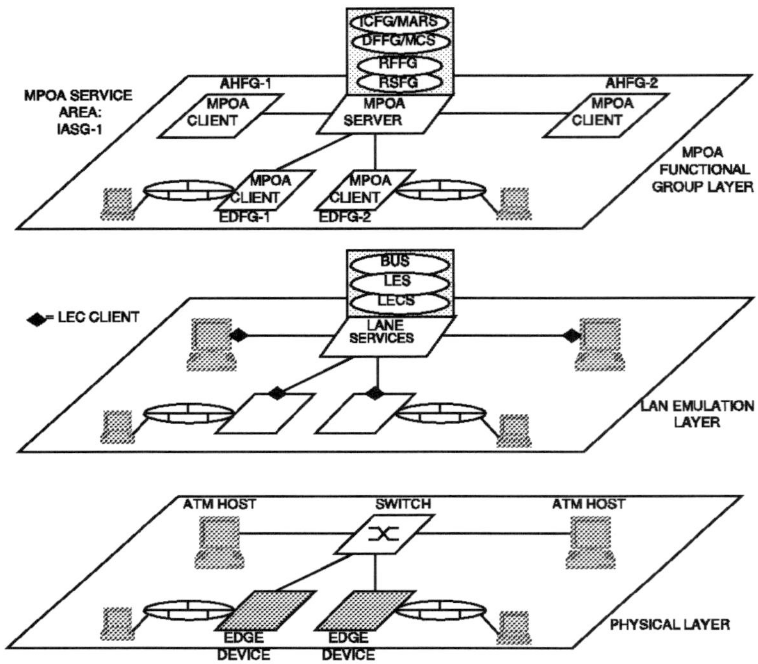

Figure 30. MPOA Functional Components

Figure 30 shows the mapping between the logical and physical components, which are split between the following layers:

- MPOA Functional Group Layer
- LAN Emulation Layer
- Physical Layer

The MPOA solution will be implemented into various functional groups that include:

Internetwork Address Sub-Group (IASG)

A range of internetwork layer addresses (for example, an IPv4 subnet). Hence, if a host operates two internetwork layer protocols, it will be a member of, at least, two IASGs.

Edge Device Functional Group (EDFG)

EDFG is the group of functions performed by a device that provides internetworking level connections between a legacy subnetwork and ATM.

- An EDFG implements layer 3 packet forwarding, but does not execute any routing protocols (executed in the RSFG).
- Two types of EDFG are allowed: *simple* and *smart*.
  - Smart EDFGs request resolution of internetwork addresses (that is, it will send a query ARP type frame if it doesn't have an entry for the destination).
  - Simple EDFGs will send a frame via a default class to a default destination if no entry exists.
- A coresident proxy LEC function is required.

ATM-Attached Host Functional Group (AHFG)

AHFG is the group of functions performed by an ATM-attached host that is participating in the MPOA network.

- A coresident proxy LEC function is optional.

  Within an IASG, LAN emulation is used as a transport mechanism to either legacy devices or LAN emulation devices, in which case access to a LEC is required. If the AHFG will not be communicating with LANE or legacy devices, then a coresident LEC is not required.

IASG Coordination Functional Group (ICFG)[6]

ICFG is the group of functions performed to coordinate the distribution of a single IASG across multiple legacy LAN ports on one or more EDFG and/or ATM device. The ICFG tracks the location of the functional components so that it is able to respond to queries for layer 3 addresses.

Default Forwarder Function Group (DFFG)[6]

In the absence of direct client-to-client connectivity, the DFFG provides default forwarding for traffic destined either within or outside the IASG.

- Provides internetwork layer multicast forwarding in an IASG; that is, the DFFG acts as the multicast server (MCS) in an MPOA-based MARS implementation.
- Provides *proxy* LAN emulation function for AHFGs (that is, for AHFGs that don't have a LANE client) to enable AHFGs to send/receive traffic with legacy-attached systems.

Route Server Functional Group (RSFG)[7]

RSFG performs internetworking level functions in an MPOA network. This includes:

- Running conventional internetworking routing protocols (for example, OSPF, RIP and BGP)
- Providing address resolution between IASGs, handling requests and building responses

Remote Forwarder Functional Group (RFFG)[7]

RFFG is the group of functions performed in association with forwarding traffic from a source to a destination, where these can be either an IASG or an MPOA client. An RFFG is synonymous with the *default router* function of a typical IPv4 subnet.

Note: One or more of these functional groups may coreside in the same physical entity. MPOA allows arbitrary physical locations of these groups.

3.4 Information Flows in the MPOA Solution

Within the MPOA network, various information flows will be present, which can be categorized as follows:

Configuration

Used by all functional groups to retrieve configuration information.

Data Transfer

The end goal of the system. The detailed formats of user data are specified as part of the internetworking protocol used and thus need not be specified as part of the MPOA solution.

Client-to-Server Control

Used by clients to query and update MPOA servers.

Server-to-Server

Used by servers to provide a single-system image, while enabling the functions to be distributed across multiple platforms for reasons of capacity and/or availability.

---

[6] ICFG/DFFG functional groups are coresident hence no protocol exists, or needs to be defined between the two.

[7] RSFG/RFFG functional groups are coresident hence no protocol exists, or needs to be defined between the two.

The result of these protocols is to enable MPOA clients to set up direct connections with each other across the ATM network based on layer 3 addresses. To enable the setup of such connections, the MPOA clients require two pieces of information:

Next hop layer 3 reachability information
: To enable determination of the layer 3 address of the MPOA client, which either supports the destination layer 3 address or through which the destination layer 3 address is reachable. This is the role of NHRP (see 5.2, "IP Address Resolution in ATM Networks" on page 114), and it is expected that NHRP will be used by MPOA as the layer 3 address resolution component.

ATM address resolution
: The resolved ATM address of the next hop node's layer 3 address.

Note: Since MPOA supports cut-through routes, this next hop address must be that of the *final* node on the ATM network through which the layer 3 address is reachable, and not that of an intermediate node (for example, a router).

3.4.1 Client-to-Server Flows

There are six types of client-to-server flows in the MPOA network, as follows:

RSFG Control (RSCtl)
: RSCtl information flow is used by MPOA clients to obtain information from the RSFG when trying to resolve the destination internetwork to an ATM address for inter-IASG data transfer.

ICFG Control (ICCtl)
: ICCtl information flow is used by MPOA clients to obtain information from the ICFG in support of destination resolution in the intra-IASG case and may also be used by the EDFG to register dynamically discovered legacy devices. Legacy device discovery is an optional feature of the EDFG.

Send to DFFG (DSend)
: DSend information flow is used by MPOA clients and servers to transmit data frames in the absence of a direct, client-to-client, shortcut connection and for broadcast and multicast frames.

Forward from DFFG (DForward)
: DForward information flow is used by the DFFG to forward frames to at least those MPOA clients and servers, within the IASG, addressed by the frame.

Send to RFFG (RSend)
: RSend information flow may be used by MPOA clients (and potentially the DFFG) to forward frames out of an IASG.

Forward from RFFG (RForward)
: RForward information flow is used to forward frames into an IASG in the absence of direct IASG-to-IASG connectivity.

3.4.2 Server-to-Server Flows

There are two types of server-to-server flows in the MPOA network, as follows:

RSFG-to-RSFG (RSPeer)
The RSPeer information flow is used by an RSFG to forward destination resolution queries (and to receive the response) for destinations in an IASG that are not served by the original RSFG.

ICFG-to-ICFG (ICPeer)
ICPeer information flow is used by an ICFG to distribute topology information to all of the ICFGs that serve a given IASG. For each ICFG, this flow can be either a single point-to-multipoint or collection of point-to-point VCCs.

Figure 31 shows some intra-IASG flows, whereas Figure 32 on page 73 shows some inter-IASG flows.

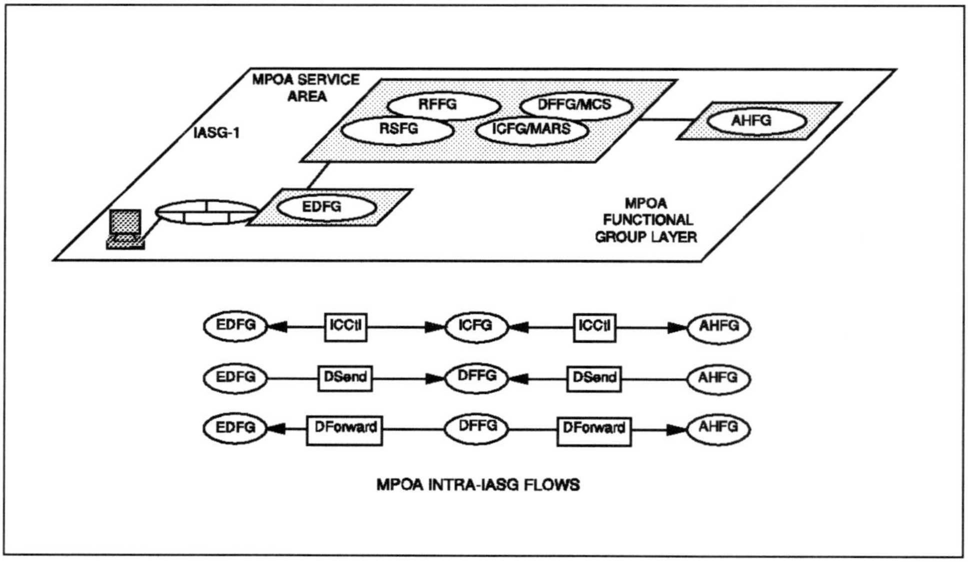

Figure 31. Intra-IASG MPOA Flows

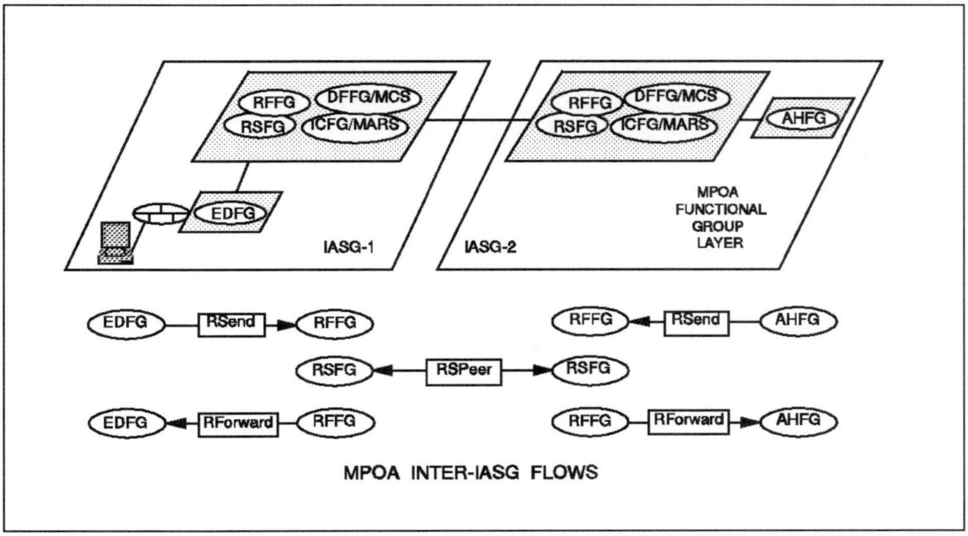

Figure 32. Inter-IASG MPOA Flows

3.4.3 Data Encapsulation

All information transferred between functional groups in the MPOA solution is transferred using the LLC/SNAP encapsulation header defined in RFC 1483 (see Appendix E, "Multiprotocol Encapsulation over AAL 5 (RFC 1483)" on page 207 and 7 on page 229).[8] The LLC/SNAP encapsulation header permits sharing of virtual circuits between the various transfer mechanisms used by the MPOA solution.

At least two frame types will be required:

- One to indicate those frames that include a MAC header. (Such frames are considered for bridging.)
- The second to indicate those frames that do not include a MAC header. (Such frames are considered for routing.)

The bridging case is required for those frames where the required MAC header cannot be correctly constructed from the contents of the frame (for example, IPV4 ARP).

- Inter-IASG traffic is sent with an LLC/SNAP header that identifies the specific protocol (referred to as internetwork or routed format).

---

8 It has been proposed that functional groups may choose to override the default LLC/SNAP encapsulation, and use no encapsulation header at all. This would be achieved via negotiation of packet format during call setup.

- Intra-IASG traffic between ATM-attached devices (AHFGs) is sent using the internetwork format.
- LANE-based intra-IASG traffic is also sent with an LLC/SNAP header:
  - This requires a change to LAN emulation (scheduled for LANE V2.0).
  - Helps reduce the total number of VCs required.

3.5 MPOA Operation

The MPOA system operates as a set of functional groups that exchange information in order to exhibit the desired behavior. To provide an overview of the MPOA system, the behavior of the components is described in a sequence order by *significant events*:

Configuration:
Ensures that all functional groups have the appropriate set of administrative information.

Registration and Discovery:
Includes the functional groups informing each other of their existence and of the identities of attached devices and EDFGs informing the ICFG of legacy devices.

Destination Resolution:
The action of determining the route description given a destination internetwork layer address and possibly other information (for example, QOS). This is the part of the MPOA system that allows it to perform cut-through (with respect to IASG boundaries).

Data Transfer:
To get internetworking layer data from one MPOA client to another.

Intra-IASG Coordination:
The function that enables IASGs to be spread across multiple physical interfaces.

Routing Protocol Support:
Enables the MPOA system to interact with traditional internetworks.

Spanning Tree Support:
Enables the MPOA system to interact with existing extended LANs.

Replication Support:
Provides for replication of key components for reasons of capacity or resilience.

Figure 33 on page 75 and Figure 34 on page 76 identify the difference in operation between route setup and data transfer. in both a single IASG-based MPOA system (Figure 33 on page 75) and a multi-IASG-based MPOA system (Figure 34 on page 76).

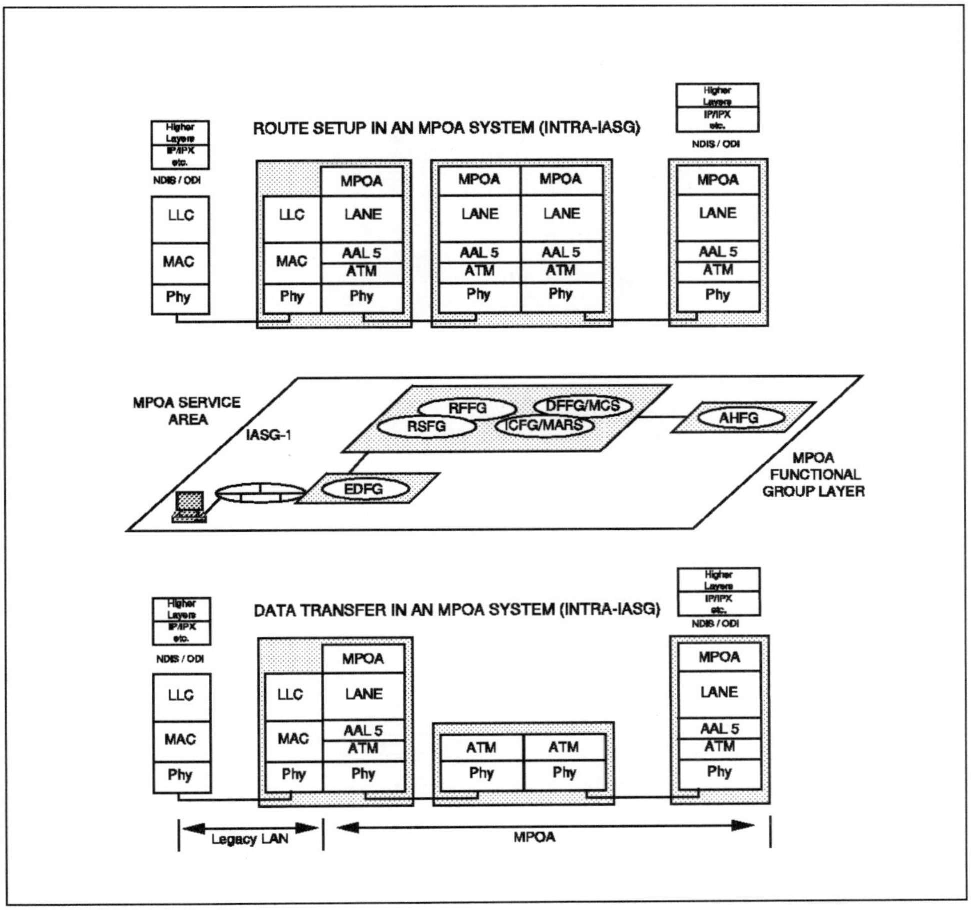

Figure 33. Route Setup and Data Transfer in a Single-IASG MPOA System

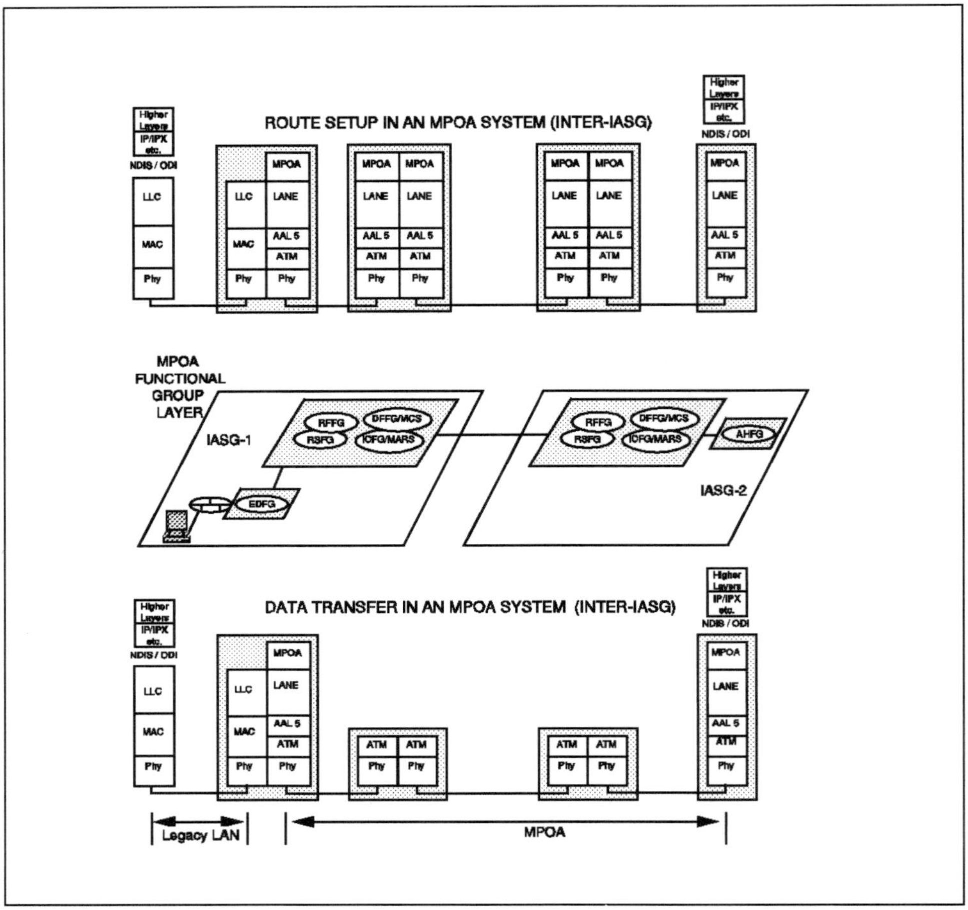

Figure 34. Route Setup and Data Transfer in a Multi-IASG MPOA System

3.5.1.1 Configuration

All functional groups will take advantage of an appropriate configuration server. The number of servers, synchronization of server databases and related tasks are not detailed here but are expected to follow the approach taken by the LAN emulation configuration service.

RSFG/RFFG Configuration

When it starts up, an RSFG/RFFG will obtain from the configuration server:

- A list of IASGs for which it is providing service, and for each:
  - The identity assigned to the IASG
  - The internetwork protocol for the IASG

- The RSFGs internetwork address on the IASG
- The information needed to join an ELAN, if any, for this IASG
- The ATM addresses of the EDFGs (if required)

- A list of routing protocols in use, and the normal configuration information required for each, including interfaces and peers, as defined by the specification of the routing protocol.

The RSFG/RFFG will establish communication with the ELAN, if any, associated with the IASG. It will then begin operating the routing protocols according to the configuration information.

ICFG/DFFG Configuration

When it starts up, an ICFG/DFFG will obtain from the configuration server:

- A list of IASGs for which it is providing coordination, and for each IASG:
  - The identity assigned to the IASG
  - The internetwork protocol for the IASG
  - Other ICFGs with which to perform coordination in support of this IASG
  - The information needed to join an ELAN, if any, for this IASG
  - The address(es) for the RDFG/RFFG pair to use for internetwork forwarding
  - The information to generate MAC addresses on behalf of AHFGs when necessary
  - The ATM addresses of the other EDFGs if required

The ICFG will establish VCs to each distinct member of the list of peer ICFGs.

Note: If the same ATM address appears in several lists of peer ICFGs, only one VC is needed between the two servers. The ICFG/DFFG will establish communication with the ELAN, if any, associated with the IASG.

EDFG Configuration

When it initializes, each EDFG will obtain from the configuration server:

- A list of IASGs of which it is to be a member, and for each IASG:
  - The IASG identity
  - The internetwork protocol supported
  - The ATM addresses of the RSFGs, RFFGs, ICFGs and DFFGs (if required)
  - The MTU for the IASG
  - The information needed to join an ELAN
- For each legacy port, a list of IASGs supported on the port
- The ELAN(s) to talk to for spanning tree propagation

The EDFG will establish RSCtl and ICCtl information flows, if appropriate, and will establish communication with the ELAN associated with each IASG. The spanning tree ELAN(s) is used to ensure that an entire MPOA family is participating in a single set of spanning tree exchanges.

AHFG Configuration

When it initializes, each host will obtain from the configuration server:

- A list of IASGs of which it is to be a member, and for each IASG:
  - The IASG identity
  - The internetwork protocol supported
  - An ICFG to use for the IASG
  - The MTU for the IASG

The host will establish VCs to each distinct ICFG. After the VCCs are set up, the host will proceed with registration.

Note: The use of a protocol-specific auto-configuration capability may require configuration information to be provided to the AHFG beyond that listed above.

3.5.1.2 Registration and Discovery

Registration in the MPOA solution is the set of exchanges that is used by the functional groups to inform each other of their existence, capabilities and interests. Discovery is the reporting by EDFGs of legacy-attached devices to the ICFG. These two mechanisms are used by the MPOA solution, along with routing protocol support, to ensure that the server components have accurate knowledge of the topology of the directly attached network, both at the MAC and internetwork levels.

RSFG Registration and Discovery

When an RSFG detects an external router, it will register the MAC address of that router with the relevant ICFG, which will in turn notify the relevant EDFG.

ICFG Registration

As the appropriate ICPeer information flow becomes available, each ICFG will register itself with each distinct peer ICFG listed in its configuration.

EDFG Registration and Discovery

When an EDFG sees legacy traffic from a previously unknown internetworking source address, on a given IASG, it may notify the ICFG.

Note: The ICFG will in turn pass on the information to other ICFGs in the same IASG.

AHFG Registration

As the appropriate VCCs become available, each AHFG will register itself with each distinct ICFG listed in its configuration. The AHFG will provide a list of IASGid, internetwork address, and optionally MAC address definitions to the ICFG. The AHFG must also inform the ICFG whether or not to serve as the AHFGs proxy on the ELAN. If the AHFG does not provide a MAC address and the ICFG has been requested to proxy for the AHFG on an ELAN, the ICFG will provide a MAC address for the AHFG.

When an ICFG receives an AHFG registration, it passes it on to all relevant ICFGs. In addition, the ICFG/DFFG will add the AHFG to all the clients DForward flows for the IASG.

3.5.1.3 Destination Resolution

When an AHFG wishes to send traffic to an internetwork destination, in both the intra and inter-IASG cases, it needs to determine the ATM address to use. This is done by sending a query to the ICFG. Upon resolving the query, the ICFG responds to the AHFG with the ATM address. If the AHFG uses mask and match behavior, it may send appropriate queries directly to an RSFG.

When an EDFG receives a local address resolution request (for example, an IPV4 ARP) on a traditional shared medium, it may determine whether it already has a cached answer for the IASG. If it does and the destination is behind another EDFG, it forwards the message to the appropriate destination. If the destination is an AHFG, the EDFG may simply create the proper address resolution response with the MAC address of the AHFG. If it does not have the information, the EDFG forwards the frame to the emulated LAN so that the frame can be forwarded to all other elements in the IASG.

3.5.1.4 Data Transfer

Unicast data flow through the MPOA system has two primary modes of operation:

- The default flow
- The shortcut flow

Shortcut flows are established by the cache management mechanisms. The default flow mechanism provides for data forwarding whenever shortcuts do not exist.

Default Unicast Data Transfer

In the default situation, the data frame is sent by an MPOA client to an RFFG or DFFG and then forwarded toward the final destination.

When an AHFG originates the packet, it may detect and choose, for internetworking destinations outside the IASG, to send the packet to the RFFG. Otherwise, the AHFG sends the packet to the DFFG.

When an EDFG sends the packet in the default situation, it uses LANE for delivery based on the MAC address in the packet.

For a dual mode host originating data, it will use LANE for transmission of intra-IASG traffic and use the AHFG for inter-IASG traffic.

When a packet arrives at an RFFG for an inter-IASG destination, it is forwarded by the routing system. When a packet arrives at a DFFG for a registered AHFG, the internetwork protocol packet is forwarded directly to the AHFG.

When a packet arrives at a DFFG from an AHFG for an intra-IASG destination that is not an AHFG, the DFFG will add the appropriate MAC header and use LANE for further forwarding. If the DFFG does not have enough information to build the MAC header, appropriate address resolution is used to get the MAC information.

When a packet arrives at a DFFG for a destination that is not within the IASG, the packet is forwarded to an RFFG.

Short Cut Unicast Data Transfer

When an MPOA client has an internetwork protocol packet to send, the packet is sent over the shortcut VC (if it exists) with the appropriate internetworking encapsulation.

Broadcast/Multicast Data Transfer

When an AHFG wishes to send a broadcast or multicast frame, it uses the ICFG in a manner similar to the above. Specifically, it sends a query to the ICFG and gets an ATM address.

Inter-IASG broadcast/multicast transfer will presumably be done either by selective splicing onto the point-to-multipoint connection that will be used by the above, or by having another kind of server within the IASG accept the information and forward it to a different point-to-multipoint circuit for inter-IASG transfer.

For EDFGs forwarding broadcast/multicast, the same set of transmissions is required.

3.5.1.5 Intra-IASG Coordination

Each ICFG is given a list of all other ICFGs supporting a given IASG. It establishes a point-to-multipoint VC to them, or uses a collection of point-to-point VCs. In either case, whenever an ICFG accepts a registration, it sends the information to all other relevant ICFGs. When a registration is removed (for example, by timeout or VC termination) all other relevant ICFGs are notified of that fact as well. Additionally, port status changes are passed on.

When an ICFG gets a query from an AHFG for a destination not within the IASG, it will pass the query to an RSFG registered as serving the source IASG.

When an ICFG receives a legacy station registration from an EDFG, it passes that information to all registered relevant RSFGs, so as to avoid the need to query the ICFG.

3.5.1.6 Routing Protocol Support

The RSFG is involved in the operation of traditional routing protocols. These may include operation-over-legacy media reached through EDFG, operation-over-ATM with identified peers, and operation-over-ATM groups using a mesh of point-to-multipoint VCs.

EDFGs may and ICFGs will send queries to an RSFG. If the destination is in another IASG handled by the RSFG, it will generate the response and record the fact of the response. If the information is later invalidated (for example, due to a routing change) a notification will go to the source of the query. This is required to avoid routing loops in the presence of routing information.

3.5.1.7 Spanning Tree Support

Each EDFG must run IEEE 802.1(d) spanning tree on each port. There is a specific IASG only for propagating spanning tree BPDUs among the EDFGs. Each EDFG is responsible for the generation of BPDUs into that IASG, and onto its attached legacy shared media.

Each EDFG will also notify the ICFG when any of its ports enter the blocking or forwarding state.

Note: This includes the ATM port which can become blocked. The blocking notification is critical since it results in invalidating routing/location information. The ATM port blocking state also interacts with *knowing* where things are, since other EDFGs are not directed to send intra-IASG traffic to that EDFG. The forwarding state notification serves to counterbalance the blocking state notification (thus avoiding *assumptions*).

3.5.1.8 Multicast Support

Multicasting is the process whereby a source host or protocol entity sends a packet to multiple destinations simultaneously using a single, local transmit operation. The more familiar cases of unicasting and broadcasting may be considered to be special cases of multicasting (with the packet delivered to one destination, or *all* destinations, respectively).

Note: In practice, the definition of broadcast transmissions is supplemented by a statement defining the scope of *all destinations*. With ATM networks it can be foolish to define *all destinations* to cover every ATM-attached interface. In practice the interfaces attached to an ATM network are usually already divided into logical or virtual layer 3 subnetworks for unicast purposes, providing a reasonable basis on which to define the scope of broadcast.

With RFC 1483 7 on page 229, the IETF defined a multiprotocol mechanism for encapsulating and transmitting packets using AAL5 over ATM virtual channels (VCs).

However, the ATM Forum's currently published signalling specifications UNI 3.0 and UNI 3.1 9 on page 230 does not provide the required multicast address function.

The most fundamental limitations of UNI 3.0/3.1's multicast support are:

- Only point-to-multipoint, unidirectional VCs may be established.
- Only the root (source) node of a given VC may add or remove leaf nodes.

Therefore, a sender must have prior knowledge of each intended recipient, and explicitly establish a VC with itself as the root node and the recipients as the leaf nodes.

Hence, to satisfy the requirement for multicast and broadcast services that layer 3 data protocols will have, we must:

- Define a group address registration and membership distribution mechanism that allows UNI 3.0/3.1-based networks to support the multicast service of protocols such as IP.
- Define specific end-point behaviors for managing point-to-multipoint VCs to achieve multicasting of layer 3 packets.

Note: It is understood that a multicast function will be inherent in UNI 4.0. UNI 3.0 and UNI 3.1 use a *root-controlled* model of point-to-multipoint VCs, and have no multicast address.

It is apparent that MPOA will have to address the need for multicast and broadcast services by layer 3 data protocols.

If we define a *cluster* as "the set of ATM interfaces that are willing and able to participate in direct ATM connections to achieve multicasting and broadcasting of AAL SDUs between themselves," a number of questions are raised:

- How do we achieve efficient multicasting of AAL SDUs around a cluster (or intracluster)?
- How will we interact with our unicast solutions (should the unicast solution simply be a subset of the general multicast solution)?

  The MPOA unicast and broadcast should be built around a core, general multicast mechanism. However, most people are already further down the development path for unicast-only solutions.

- What minimal UNI functionality will MPOA expect to work with?

  If we wish for MPOA to have a solution useful to the present and growing base of installed ATM networks, we must support intracluster multicasting using UNI 3.1.

Chapter 4. APPN Support in ATM Networks

IBM, in conjunction with the APPN Implementers Workshop (AIW), is currently working on enhancements to APPN architecture that describes a native ATM DLC that APPN nodes can use to gain access to ATM networks. These enhancements will allow existing APPN applications to gain access to ATM QOS and traffic contracts without changes being made to the applications themselves. In addition, a method where other protocols can be transported on the same ATM virtual circuit as APPN data is also defined. This method does not involve encapsulation of one protocol inside another, but allows true multiplexing on a single ATM VC. Native access to ATM networks will allow existing APPN to use and gain the full benefits of ATM without the use of an enabling protocol, for example, multiprotocol over ATM (MPOA).

The scope of these enhancements is limited to APPN products. IBM's current strategy is for subarea products to gain access to ATM either through migration to APPN or through LAN emulation or frame relay interworking.

High performance routing (HPR), as an extension to base APPN, was standardized in 1995 by the AIW. HPR implementations are available from IBM and other AIW members. For a more detailed description of APPN and HPR, see the *APPN Architecture and Product Implementations Tutorial*, GG24-3669 (refer to 18 on page 230).

The information in this chapter is intended to give an overview of the design work that has been done on a native ATM DLC for APPN. This information may change before the architecture is accepted by the AIW.

4.1 High Performance Routing (HPR)

High performance routing (HPR) enhances APPN data routing performance, especially when using high-speed links. HPR's nondisruptive path switching can also make APPN more reliable by routing APPN sessions around network outages. HPR allows switching to be done at a lower layer, and therefore much faster, in intermediate nodes along an APPN session path. HPR changes the existing intermediate session routing (ISR) of APPN by using a routing algorithm, which minimizes the storage and processing requirements in intermediate nodes. The level of error recovery done in base APPN is no longer necessary for today's more reliable high-speed lines. Instead of error recovery on individual lines, HPR provides an end-to-end level of error recovery. HPR also enhances APPN by providing a nondisruptive path switch function that can route sessions around failed links or nodes. APPN flow control is done on each stage of a session, using adaptive session pacing. Adaptive session pacing works well in a network comprised of a mixture of link-types, with differing speeds and quality. However, for high-speed networks, adaptive session pacing is not adequate because of the amount of processing required in the intermediate nodes. Figure 44 on page 104 shows how automatic network routing (ANR) has replaced ISR, and how ANR will be replaced by ATM switching in ATM networks.

The two main components of HPR are the rapid-transport protocol (RTP) and automatic network routing (ANR).

Rapid-Transport Protocol

RTP is a connection-oriented full-duplex protocol designed to support data in high-speed networks. RTP connections can be thought of as *transport pipes* through an HPR subnet.

The RTP functions include:

Nondisruptive path switch (NDPS)

An RTP connection's physical path can be switched automatically to reroute around a failure in the network. Any data that was in the network at the time of failure will be recovered automatically by RTP's end-to-end error recovery.

End-to-end error recovery

In base APPN, error recovery is done separately on every link in the network. To address the needs of high-speed lines with lower error rates, HPR does error recovery on an end-to-end basis. RTP also supports selective retransmission, where only missing or corrupted packets are resent, and not all packets since the error occurred.

End-to-end flow and congestion control

RTP provides a new method called adaptive rate-based congestion control (ARB). ARB regulates the flow of traffic by predicting network congestion and reducing a node's sending rate into the network. Thus ARB will prevent congestion rather than reacting to it once it has occurred. ARB allows networks to be designed with higher link utilizations.

Automatic Network Routing

Automatic network routing (ANR) is a routing mechanism to minimize storage and processing requirements for routing packets through intermediate nodes.

The ANR functions include:

Fast packet switching

ANR takes place at a lower level than ISR. Functions such as link-level error recovery, segmentation, flow control and congestion control are no longer performed in the intermediate nodes. These functions are now performed at the RTP connection endpoints.

No session awareness

Intermediate nodes are not aware of the SNA sessions or the RTP connections that are established across the node. This means there are no requirements to keep routing tables for session connectors; these are needed in base APPN, and need between 200 and 300 bytes per session per node. This storage saving will be essential when HPR nodes supporting high-speed links will be carrying many more intermediate sessions than APPN nodes do today.

Source routing

ANR is a source-routing protocol that carries the routing information in a network header with the packet. Each node strips off the information it has used in the packet header so that the next node can easily find its routing information at a fixed place in the header. This means that switching packets through a node can be done more quickly than in the routing table lookup method used in base APPN.

Error Recovery

Base APPN traffic uses LLC 802.2 type 2 connections, which can provide error-recovery on LANs. HPR traffic is optimized to use links with low error rates. On such links, end-to-end error recovery is performed by RTP instead of hop-by-hop error recovery performed by LLC (shown in Figure 44 on page 104).

4.2 Techniques for ATM Utilization

There are currently three methods by which APPN could utilize ATM:

- LAN emulation
- Frame relay interworking
- APPN native ATM DLC

The first two discussed, LAN emulation and frame relay interworking, are possible without changes being made to the higher-layer software. The third method, ATM native DLC, involves changes, but allows APPN to make full use of ATM services.

4.2.1 LAN Emulation

The goal of LAN emulation (LANE) is to allow higher-layer protocols (for example, TCP/IP or APPN) to be unaware that their data is traversing an ATM network. LAN emulation enables layer 3 protocols to access an ATM network as if they were running over a so-called legacy LAN. As a result, no changes are needed to the higher-layer protocols. For more details on LANE, see 2.1, "LAN Emulation Version 1.0" on page 45.

LAN emulation software does not allow the higher layers to specify any ATM specifics, such as quality of service (QOS) or throughput parameters. The higher layer is dependent upon the static choice for ATM parameters made by the LAN emulation software. Generally, LAN emulation software uses ATM signalling to set up ATM virtual connections with null SSCS, the common part convergence sublayer (CPCS) of AAL type 5, the segmentation and reassembly sublayer (SAR) of AAL type 5 and, typically, best-effort service.

Frame Relay Interworking: In the near future, many customers may not be able to afford end-to-end ATM technology deployment. Frame relay technology often requires only an upgrade to existing technology. As a result, frame relay equipment is already widely deployed. For data transfer, frame relay provides the needed services and is a good first step into the switched environment. There are two techniques that allow interworking between frame relay terminating equipment and ATM terminating equipment.

Network Interworking

Where LAN emulation makes an ATM network look like a legacy LAN to the higher-layer protocols, network interworking makes an ATM network look like a frame relay bearer service.

Network interworking is an encapsulation technique in which frame relay packets are segmented into ATM cells. Network interworking is currently defined only for frame relay permanent virtual circuits, therefore there is no signalling translation defined.

Service Interworking

The second technique that allows interworking between frame relay and ATM networks is called *service interworking*. Service interworking requires a translational gateway to transform frame relay packets into ATM cells and vice versa. Translation is required between the two multiprotocol encapsulation methods, RFC 1483 for ATM and RFC 1490 for frame relay. Service interworking is currently defined only for frame relay PVCs, as signalling translation between the two technologies would be required. Based on the complexity of this translation, it is not clear when service interworking will be defined for SVCs.

Service interworking requires that both nodes have a compatible logical link control function. LDLC is the base for ATM, and IEEE 802.2 type 2 (LLC2) is the base for frame relay. These LLCs use different mechanisms to determine when an activation XID exchange is complete, to deactivate a TG, and to monitor link availability. For service interworking to work for HPR traffic, either optional frame relay LDLC support must be implemented by an HPR node on the frame relay side, or optional LLC2 support must be implemented by the HPR node on the ATM side. For service interworking to work for FID2 traffic, ATM LLC2 support must be implemented by the HPR node on the ATM side.

4.2.2 Native ATM DLC

The development of an ATM DLC is a more straightforward approach to APPN utilization of ATM. This approach eliminates the indirection and restrictions of the LAN emulation approach. Native ATM DLCs require changes to the higher-layer protocol software (for example, to accept ATM addresses at the MAC driver interface).

The following prerequisites have been defined for APPN communication over native ATM DLCs:

- The base functions for APPN architecture Version 2 (see 20 on page 231).
- High performance routing enhancements including the Rapid Transport Protocol (RTP) and control flows over RTP option sets.

The decision to use HPR was made because the *go-back-n* error recovery mechanism used by 802.2 type 2 LLC is not sufficient for high-speed ATM links. Instead *selective retransmission* is needed. Selective retransmission can be provided by RTP or an LLC like the service specific connection-oriented protocol (SSCOP). The HPR prerequisite eliminates the need to implement a high-function LLC. As a result, 802.2 type 1 LLC can be used for ATM, and a new logical data link control (LDLC) has been designed to provide functions like reliable delivery of XIDs.

Unlike LAN emulation, a native ATM DLC allows APPN to fully exploit ATM's guaranteed bandwidth services. Frame relay interworking can provide similar services, but only on a subscription basis until frame relay SVC interworking is defined. In addition, a native ATM DLC would allow APPN to exploit ATM services for real-time transport and multicast, functions that are not provided by a frame relay service. The AIW proposal for a native APPN DLC does not support real-time traffic or multicast at this time.

For time-critical transactions, reserved-bandwidth variable bit rate connections with controlled delay and error rates could be allocated, whereas batch file transfers could use the cheaper less predictable unspecified bit rate (UBR) connections. SNA allows HPR to match an ATM connection's quality of service (QOS) to the COS needs of an application.

4.3 Native ATM DLC Implementation

When APPN/HPR runs over an ATM DLC, it is better able to exploit the features of ATM, such as its quality of service (QOS), but there are many considerations (for example, which AAL type to use for the user plane) that must be taken into account and APPN enhancements required to allow exploitation. The following sections describe these considerations and enhancements.

4.3.1 Node Structure

Figure 35 on page 89 shows the node structure for a node supporting only APPN/HPR traffic; products will be free to implement other structures (if needed). The node structure for support of other protocols (for example, IP) has also been defined. The native ATM DLC includes the ATM signalling and LDLC components. The ATM signalling component converts configuration services (CS) signals into the signals defined on the interface to the Port Connection Manager (PCM) (typically located on the ATM adapter) and vice versa. The low-level ATM interface (LL ATMI) defines such an interface. Reliable delivery is provided by logical data link control (LDLC) for a small set of APPN flows (XID, XID_DONE and DEACT). Error recovery for HPR RTP packets is provided by the protocols at the RTP endpoints.

LDLC, using the HPR network header, multiplexes traffic from CS with HPR RTP traffic. The HPR network control layer (NCL) uses the automatic network routing (ANR) information in the HPR network header to pass incoming packets to either the RTP or to an outgoing link. RFC 1483 (see Appendix E, "Multiprotocol Encapsulation over AAL 5 (RFC 1483)" on page 207) defines multiprotocol encapsulation over ATM; it provides for encapsulation of HPR NLP packets within 802.2 type 1 headers.

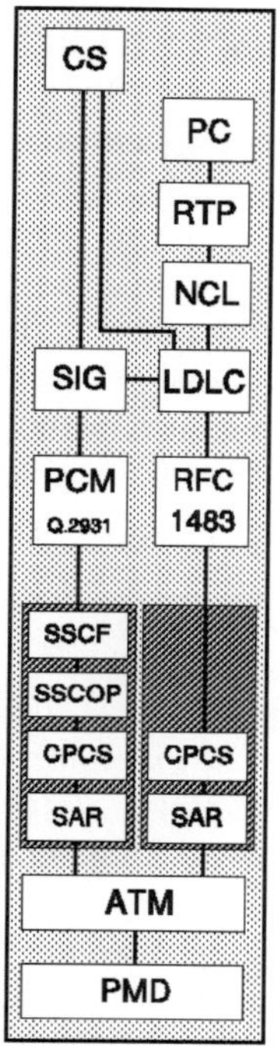

Legend:

- Control Plane
  - CS Configuration Services
  - SIG ATM Signalling
  - PCM Port Connection Manager
  - SAAL Signalling ATM Adaptation Layer
    - SSCF Service Specific Coordination Function
    - SSCOP Service Specific Connection Oriented Protocol
    - CPCS Common Part Convergence Sublayer
    - SAR Segmentation and reassembly Sublayer
- User Plane
  - PC Path Control
  - RTP Rapid Transport Protocol
  - NCL Network Control Layer
  - LDLC Logical Data Link Control
  - RFC 1483
  - AAL5 ATM Adaptation Layer 5
    - CPCS, SAR
- ATM ATM Layer
- PMD Physical Medium Dependent Layer

Figure 35. *Node Structure*

Products will be free to implement other structures if needed.

4.3.1.1 Low Level ATM Interface (LL ATMI)

The native ATM DLC approach requires an interface to ATM that gives higher-layer protocols the ability to request the full range of ATM services. APPN products are free to implement any such interface.

IBM has proposed the *low level ATM interface* to the ATM Desktop Alliance. LL ATMI provides a common semantics platform for access to the native services of ATM as defined by the ATM Forum's *User-to-Network Interface Specification, Version 3.1.* (see 9 on page 230). The specification (file name ATMIXPS.ZIP) is available via anonymous FTP at ftp://ftp.efficient.com/pub/desktopapi.

The LL ATMI is an interface between the network adapter driver and the higher-layer protocols (perhaps using a common connection manager). The interface provides a formal description of the interface semantics and operating system-independent message encodings. This interface is for both the signalling and user planes. The LL ATMI's positioning is platform-specific for both the user and control planes. For signalling, it is typically positioned between the higher-layer protocols or their common connection manager and the Q.2931 port connection manager. For the node structure in Figure 35 on page 89, it is typically positioned immediately above the AAL 5 for the user plane.

The LL ATMI should not be confused with a UNI, which defines the interface between an ATM end station and the public or private ATM network equipment over which signalling information and data are exchanged. The LL ATMI is an interface within an ATM end station over which the higher layers gain access to ATM services.

4.3.1.2 Control Plane

In order for an APPN node to dynamically establish, maintain and terminate SVC connections through an ATM network, the node uses ATM signalling procedures to exchange information (for example, which ATM adaptation layer type will be used for the SVC) with the network. Depending on whether the network is private or public, the interface is referred to as either a private UNI or public UNI. UNI signalling is standardized by the ATM Forum in ATM UNI 3.1 and by the ITU. Q.2931 is the layer 3 protocol used to control the UNI. The component providing Q.2931 signalling is called the port connection manager (PCM). Q.2931 runs on top of the signalling ATM adaptation layer (SAAL), which defines how to transfer the signalling information reliably using cells of the ATM layer on virtual channels. This is described in ITU-T recommendation Q.2100, B-ISDN Signalling ATM Adaptation Layer Overview Description. Currently a dedicated point-to-point signalling virtual channel with VCI=5 and VPI=0 is used for UNI signalling.

SAAL consists of a service-specific part and a common part. The service-specific part further consists of a UNI service-specific coordination function (SSCF) and a service specific connection-oriented protocol (SSCOP). The UNI SSCF maps the particular requirements of Q.2931 to the requirements of the ATM layer. This is defined in ITU-T recommendation Q.2130, B-ISDN ATM Adaptation Service Specific Coordination Function for Signalling at the User-to-Network Interface. SSCOP provides mechanisms for the establishment, release and monitoring of signalling information exchange connections between peer signalling entities. This is described in ITU-T recommendation Q.2110, B-ISDN ATM Adaptation Layer Service Specific Connection-Oriented Protocol. SAAL uses the common part convergence sublayer (CPCS) and the segmentation and reassembly sublayer of AAL type 5.

4.3.1.3 User Plane

The ATM adaptation layer (AAL) supports higher-layer functions of both the user and control planes. The SAAL, described earlier, is used for the control plane.

There are several AAL types defined for the user plane. AAL type 3/4 and AAL type 5 are used for variable bit rate (VBR) data. The AAL type used for a given SVC is defined with the signalling protocols in the AAL information element. The structure for the user plane is shown in Figure 35 on page 89.

The CPCS performs functions common to all AAL users. The service-specific requirements of different classes of users are implemented in the service-specific convergence sublayer (SSCS). For user classes that do not require any service-specific function, the SSCS may be null. The SSCS for a given connection is specified with the signalling protocols in the ATM adaptation layer information element (IE).

The native ATM DLC for APPN uses AAL type 5 with a null SSCS.

4.3.1.4 Logical Data Link Control

LDLC performs the following functions:

Reliable delivery of XIDs
: As done by current APPN DLCs, LDLC delivers XID3s reliably.

Indication of when the XID exchange is complete
: This is analogous to the set mode function (for example, SABME and UA) and is required because configuration services (CS) needs to synchronize the completion of the XID exchange with the partner.

Deactivation of the link
: This function enables HPR to signal an adjacent node when access to a shared SVC or PVC has been terminated.

NLP Routing
: NLPs of type ANR (that carry all HPR session and control traffic) are sent and received over the link in UI frames. LDLC routes received NLPs of this type to the appropriate upper-layer component (NCL). LDLC does not guarantee successful delivery of these packets as this is provided by RTP.

Link INOP processing
: On many link types (for example, ATM and frame relay) failure notification is provided by the service provider subnet when the link connection fails. On these link types, LDLC receives an INOP message when the link connection fails. LDLC cleans up the link when an INOP is received.

Liveness Protocol
: LDLC may optionally check that the partner is alive by periodically sending *test* commands and receiving, if the partner is alive, a test response. The format of

these commands is defined in the IEEE 802.2 standard. However, this *liveness* protocol is unique to LDLC.

These functions are the only ones required since all other traffic (CP-CP session, LU-LU session, and route setup) is delivered reliably by RTP.

4.3.1.5 Error Recovery Positioning

Error recovery can be provided either by the ATM network using an SSCS or at a higher layer such as LLC. For APPN transmissions over a native DLC, error recovery will not be provided by the ATM network. This choice was made for the following reasons:

- When a VCC is established, its SSCS is specified by signalling or definition and used for all data flowing over the VCC. Thus, if the reliability mechanism is associated with that SSCS, it is the only one available for all data streams. Alternatively, with a null SSCS and reliable delivery provided at the DLC layer, each traffic stream can have its own reliable delivery mechanism (or none at all); thus, traffic streams with different reliable delivery mechanisms can be multiplexed over a single VCC.
- Multiprotocol encapsulation, as defined by RFC 1483 (and extensions), expects to run over AAL type 5 with a null SSCS (see Figure 35 on page 89). Placement of SSCOP as an SSCS would prevent interoperability with other vendor's products, which are expected to use RFC 1483 for multiprotocol data.

The rapid transport protocol (RTP) for APPN/HPR also provides error recovery and selective retransmission. Thus, RTP data does not require error recovery by LLC. To eliminate the need for a high-function LLC, HPR and RTP were made prerequisites for the native ATM DLC function. In addition, the HPR control flows over RTP tower was also made a prerequisite; thus, HPR CP-CP sessions and route setup traffic will flow only over RTP connections. Optional link-level error recovery is allowed when using LLC2 instead of LDLC over ATM's low error-rate links.

APPN/HPR requires guaranteed delivery across its links for XID3 traffic. For this reason, current DLCs used for XID3 traffic include an LLC that can provide this function. In order to provide reliable delivery for XIDs, LLC typically sends XIDs as unnumbered commands (with the poll bit set to 1) and responses. A similar technique must be provided across the user plane for ATM links. This function will be provided by the native DLC in a new logical data link control (LDLC) component. Therefore, no error recovery function need be provided by the LLC. Thus, IEEE 802.2 LLC type 1 is sufficient for native ATM.

To support frame relay service interworking, products may optionally support IEEE 802.2 LLC type 2 (LLC2). XID is used to determine whether LLC2 or LDLC will be used.

4.3.1.6 Internal Routing of Frames

When LDLC is used, APPN/HPR passes outgoing RTP traffic through its LDLC component. XID, XID_DONE, and DEACT are processed by the LDLC reliable delivery function (see Figure 36 on page 94).

The mechanism for routing frames received over an ATM network to the proper component within a node is as follows:

- All frames are encapsulated within an RFC 1483 header (see Figure 37 on page 96).
- The 1483 header indicates the higher-layer protocol to which the frame should be passed.
- When the 1483 header indicates the higher-layer protocol is HPR, the RFC 1483 header is removed, and the packet is passed to the correct instance of LDLC; that is, RFC 1483 decides which instance of LDLC is correct by looking at the SAPs in the second LLC1 header, and not the SAPs in the RFC 1483 header. LDLC examines the LLC1 header and the HPR network header.
  - When the LLC1 header indicates XID or test, or the LLC1 header indicates unnumbered information (UI) and the network header indicates function routing, the packet is processed by the LDLC reliable delivery function. LDLC forwards XID, XID_DONE and DEACT frames to APPN configuration services (CS).
  - When the LLC1 header indicates UI and the network header indicates ANR routing, the packet is passed to NCL. NCL examines the ANR information and passes the packet either to RTP or to an outgoing link.

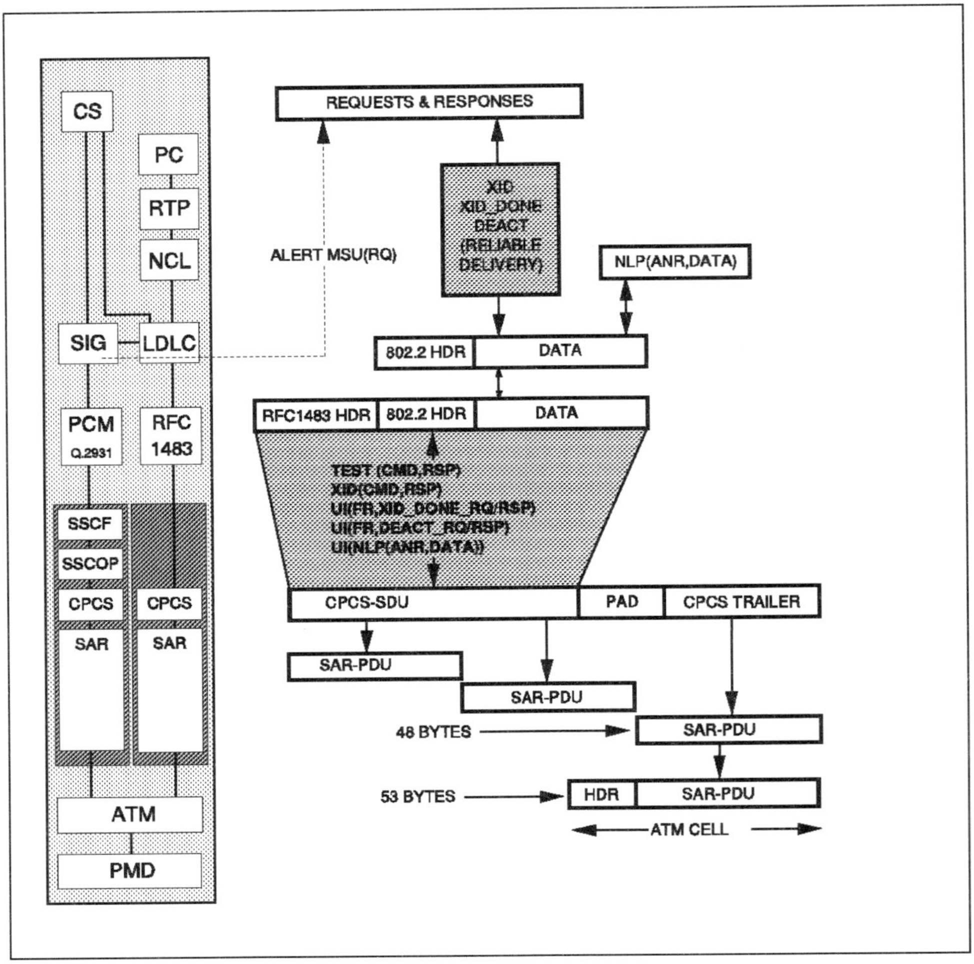

Figure 36. Internal Routing of Frames

4.3.1.7 ATM User Plane Frame Formats

The native ATM DLC will operate over AAL type 5 with a null SSCS. The network control layer of HPR will pass data network layer packets (NLP) to LDLC. How control NLPs are passed to LDLC, and internal data formats, are both implementation-dependent. Figure 36 shows a basic representation of how packets are passed down through the various layers.

The data unit passed from the RFC 1483 encapsulation function to AAL type 5 is called the CPCS service data unit (CPCS-SDU). CPCS pads the CPCS-SDU and adds an 8-byte

CPCS trailer. The resulting data unit is a multiple of 48 bytes in length and called the CPCS protocol data unit (CPCS-PDU). SAR segments the CPCS-PDU into 48-byte SAR-PDUs, which it passes to the ATM layer. The ATM layer adds its 5-byte header to each SAR-PDU to create a 53-byte ATM cell.

Figure 37 on page 96 depicts the various frame formats. All transmissions on an ATM TG will be in an IEEE 802.2 LLC frame that begins with an 8-byte header. The contents of this header are defined by RFC 1483 and ATM Forum Implementation Agreement 94-0615, which is called an RFC 1483 header. When DSAP, SSAP and Control Field are coded X'FEFE03', the fourth byte is a network layer packet identifier (NLPID). An NLPID of X'09' indicates that the NLPID is followed by a 2-byte layer 2 protocol identifier (L2) and a 2-byte layer 3 protocol identifier (L3), the format of which complies with broadband low-layer information specified in ITU-T Recommendation Q.2931. The values of L2 and L3 are defined in ATM Forum Implementation Agreement 94-0615. An L2 value of X'4C80' indicates the use of IEEE 802.2 as the L2 protocol, and an L3 value of X'7085' indicates that HPR is the layer 3 protocol.

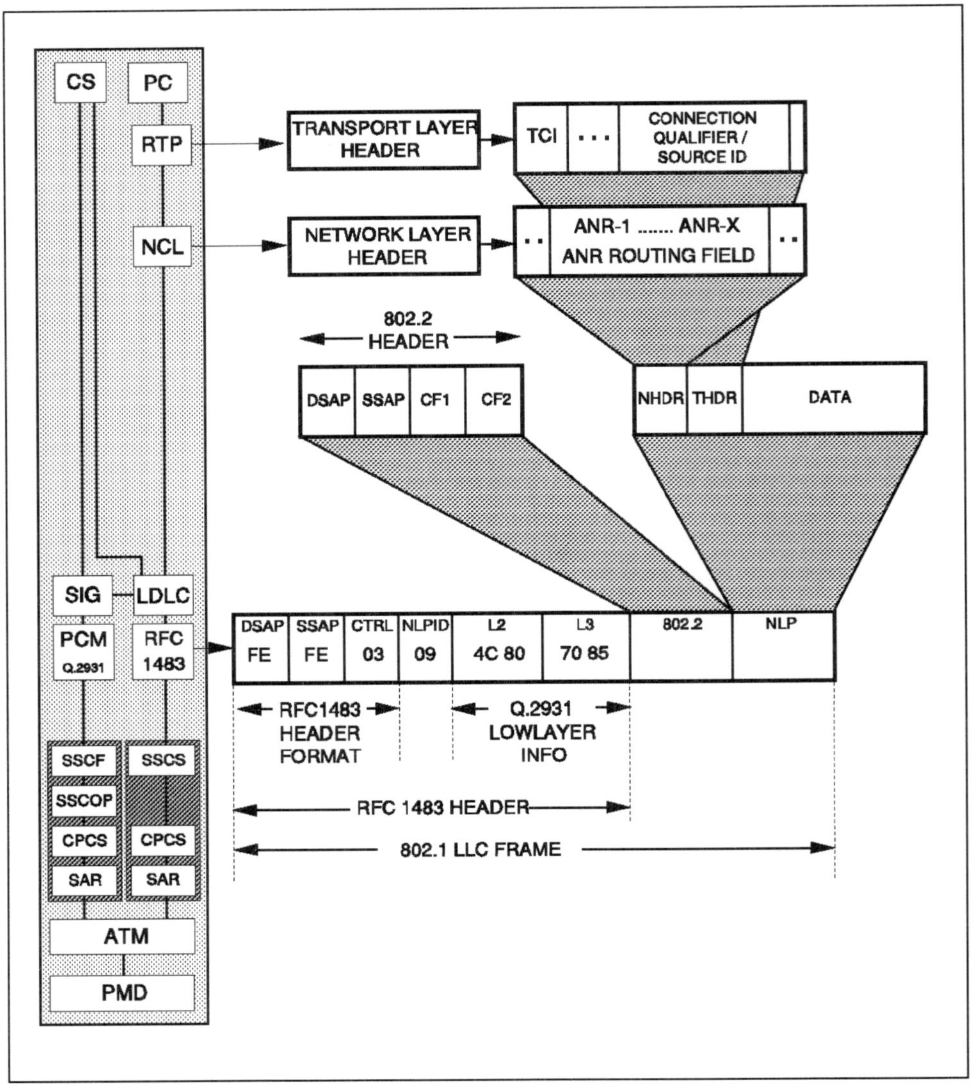

Figure 37. Frame Formats

Typical formats for data and control NLPs on a single protocol SVC are shown in Figure 38 on page 97. A value of X′101′ in the switching mode field of the network layer header (NHDR) indicates the mode is function routing. For function routing, a value of X′1′ in the function type field of the NHDR indicates that the function type is LDLC. When the function type is LDLC, there is no transport header (THDR) and a 1-byte function routing header follows the NHDR.

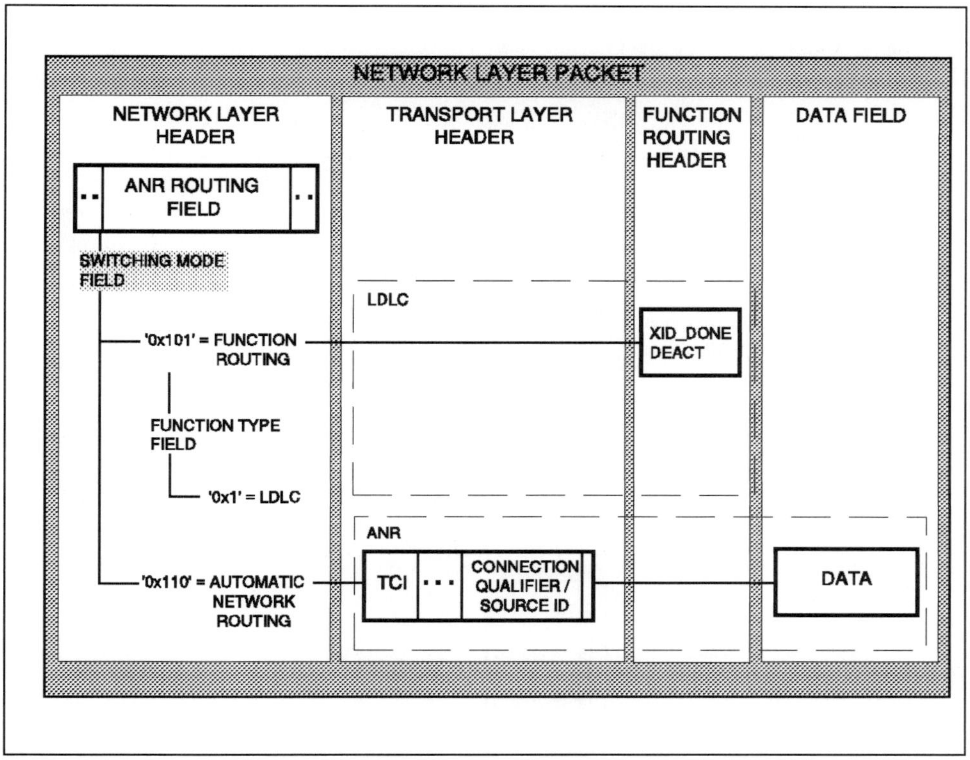

Figure 38. *NLP Formats*

4.4 ATM Connection Networks

In the connection network model, a virtual routing node (VRN) is defined to represent the shared access transport facility (SATF). Each node attached to the SATF defines a single TG to the VRN rather than TGs to all other attached nodes.

The following extensions to the connection network model for LANs are required for ATM connection networks:

- For LANs, the DLC signalling information, which consists of the MAC address and the LLC SAP address, is sufficient to establish a connection; however, this is not the case for ATM switched facilities. The DLC signalling information for ATM includes the ATM address, but other information, which may be either included in the DLC signalling information or defined locally, is required to establish a call. For example, the QOS class for the forward direction is locally defined at the node placing the call.

- The connection network model for LANs allows only one TG between a port and a VRN. For ATM, multiple TGs between a port and a VRN are allowed in order to support separation of traffic for different classes of service.
- The LAN connection network model assumes the same characteristics for each connection crossing the LAN. For ATM when multiple TGs are defined to a VRN, each may have different associated call request parameters. In addition, ATM connections across the same TG to different destination nodes may have different call request parameters based on parameter definition for the paired connection network TG.
- Normally, one connection network is defined on a LAN (that is, one VRN is defined.) For ATM, separate connection networks are required for best-effort service and reserved bandwidth connections. In addition, a separate connection network may be defined between the nodes connected to a private campus ATM network.

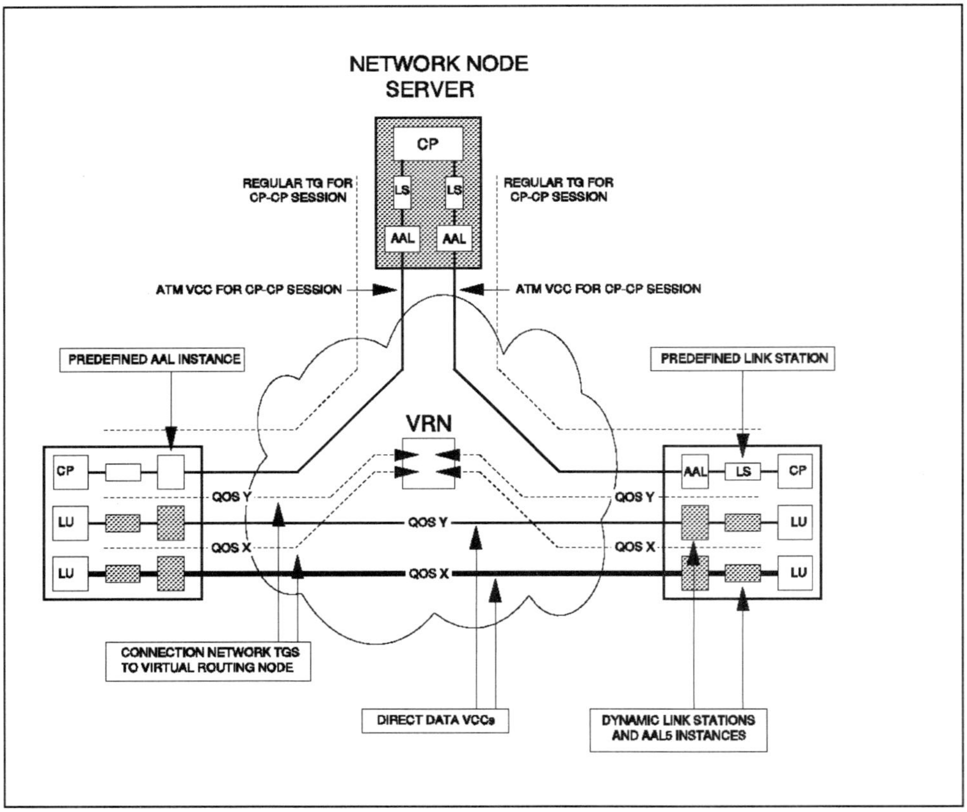

Figure 39. Connection Network Example

4.5 ATM Traffic Contracts and Quality of Service (QOS) Classes

ATM networks are expected to support a variety of data types with different characteristics. Design and operation of network control functions such as call admission, bandwidth reservation, and congestion control require accurate source characterization to achieve high resource utilization. However, some sources are unable to provide a detailed description of their traffic behavior. Hence, there is a trade-off between how much information should and can be defined to characterize a source.

The ATM UNI provides the protocol for establishing a virtual channel connection (VCC) on demand. Two traffic contracts (one for each direction) specify the negotiated throughput characteristics of an ATM connection at the UNI. The APPN node requesting the setup of the VCC selects a QOS class for each direction from the set of QOS classes supported by the ATM network. Upon agreement, the network commits to meet the requested QOS for a direction as long as the user complies with the traffic contract for that direction.

An SVC for APPN traffic needing guaranteed throughput would request specified QOS class 3, which is defined to support service class C, connection-oriented data transfer.

There is also an unspecified QOS class used with best-effort service for which no explicit characteristics are negotiated with the network. For best-effort service, there are no traffic throughput guarantees; the only parameter specified is the peak cell rate, which is used by the other endpoint.

4.6 APPN/HPR Flows over ATM

Figure 40 on page 100 and Figure 41 on page 100 show some of the basic protocol stacks involved in using the APPN native ATM DLC. All APPN - APPN communication takes place exactly as before. The native ATM DLC approach is regarded by APPN as just another DLC, although there are additions that were made for ATM signalling. Figure 40 on page 100 shows the typical protocol stacks traversed by the CP-CP sessions between APPN nodes.

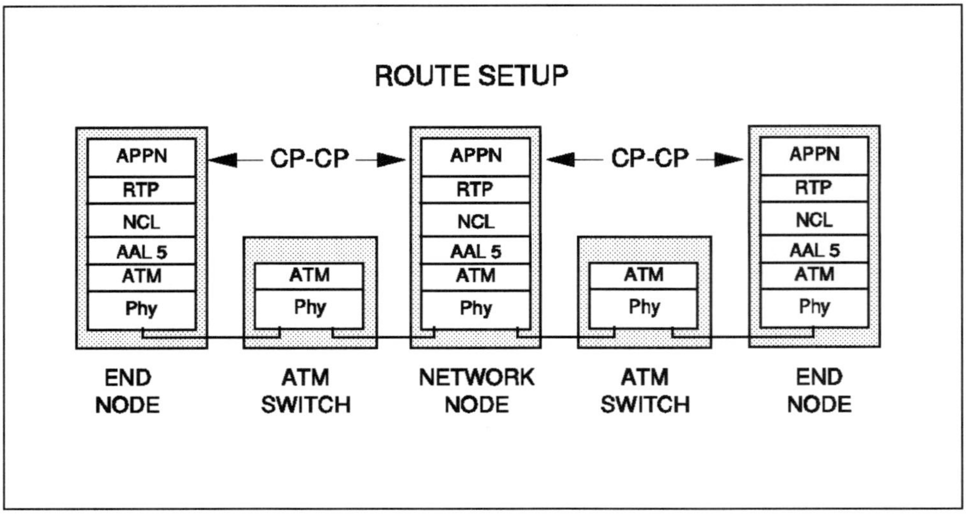

Figure 40. *APPN ATM CP-CP Sessions*

Figure 41 and Figure 42 on page 101 show the typical protocol stacks traversed by the LU-LU data session between APPN nodes.

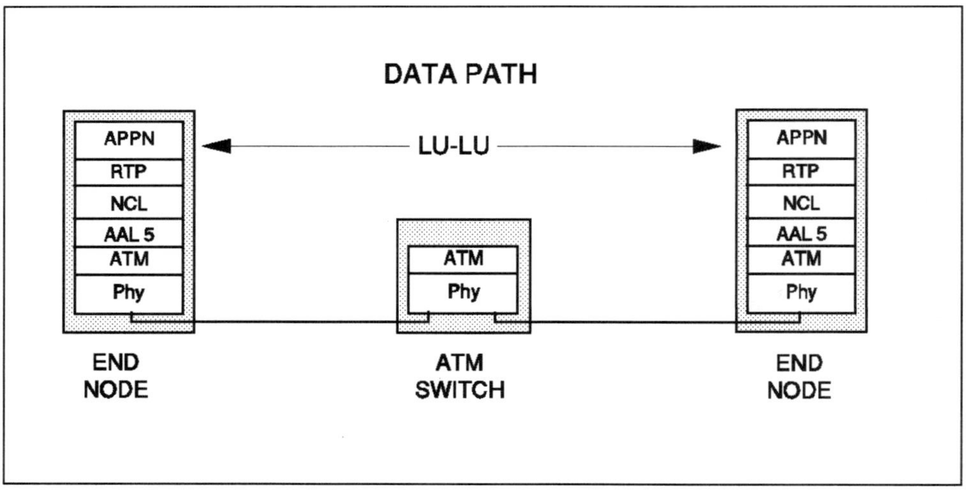

Figure 41. *APPN ATM Data Path*

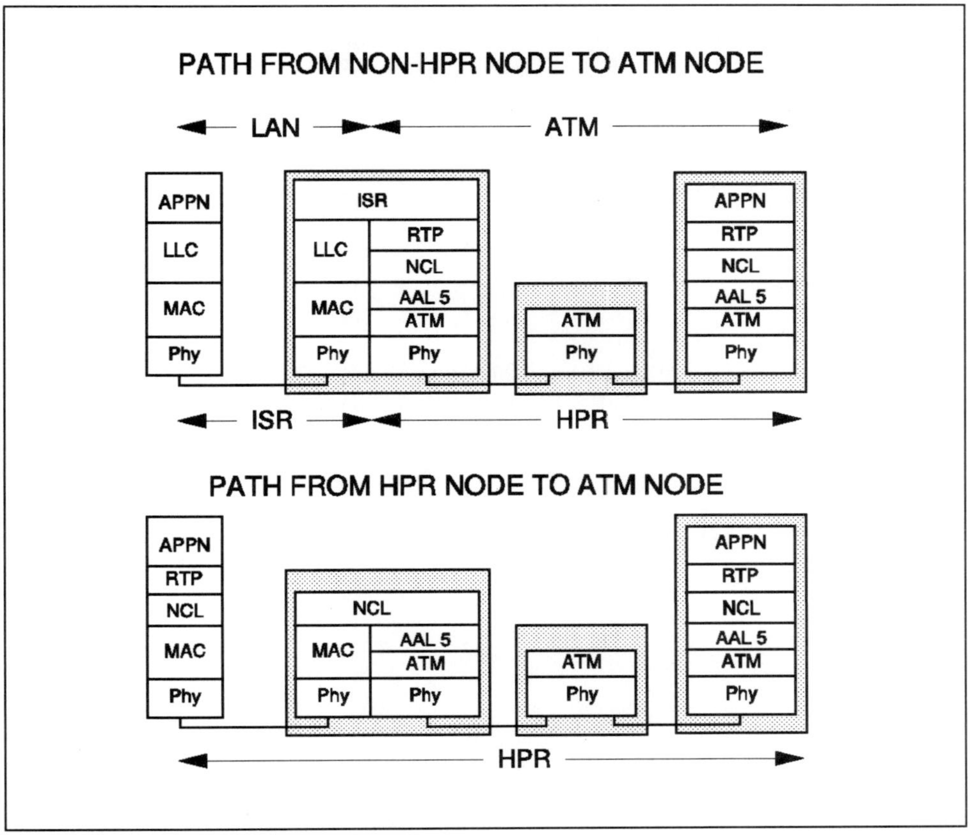

Figure 42. APPN LAN-to-ATM Data Path

4.7 Multiprotocol Encapsulation

Multiprotocol encapsulation provides a flexible method for carrying multiple protocols on a given ATM connection. The method is useful when customers desire *parallel* transport of data from multiple higher-layer protocols (that is, data from one protocol is not encapsulated within the headers of a second protocol.) Otherwise, separate VCCs must be established for each protocol. Figure 43 on page 102 shows the basic node structure for supporting multiprotocol encapsulation (in this case, IP and APPN). A common connection manager (CxM) supports signalling for the multiple higher-layer protocols.

RFC 1483, *Multiprotocol Encapsulation over ATM Adaptation Layer 5*, describes multiprotocol encapsulation for connectionless network interconnect traffic and routed and bridged protocol data units (PDUs). IBM has submitted ATM Forum contribution 94-0615, *Multiprotocol over ATM Adaptation Layer Type 5 Implementation Agreement*, that extends RFC 1483 to cover connection-oriented protocols. The implementation agreement adds code points for the following protocols:

- Subarea SNA (FID4)
- Peripheral SNA (FID2)
- APPN (FID2)
- APPN/HPR
- NetBIOS

These extensions are currently under consideration by the ATM Forum. The extensions were also presented at the APPN Implementers Workshop (AIW).

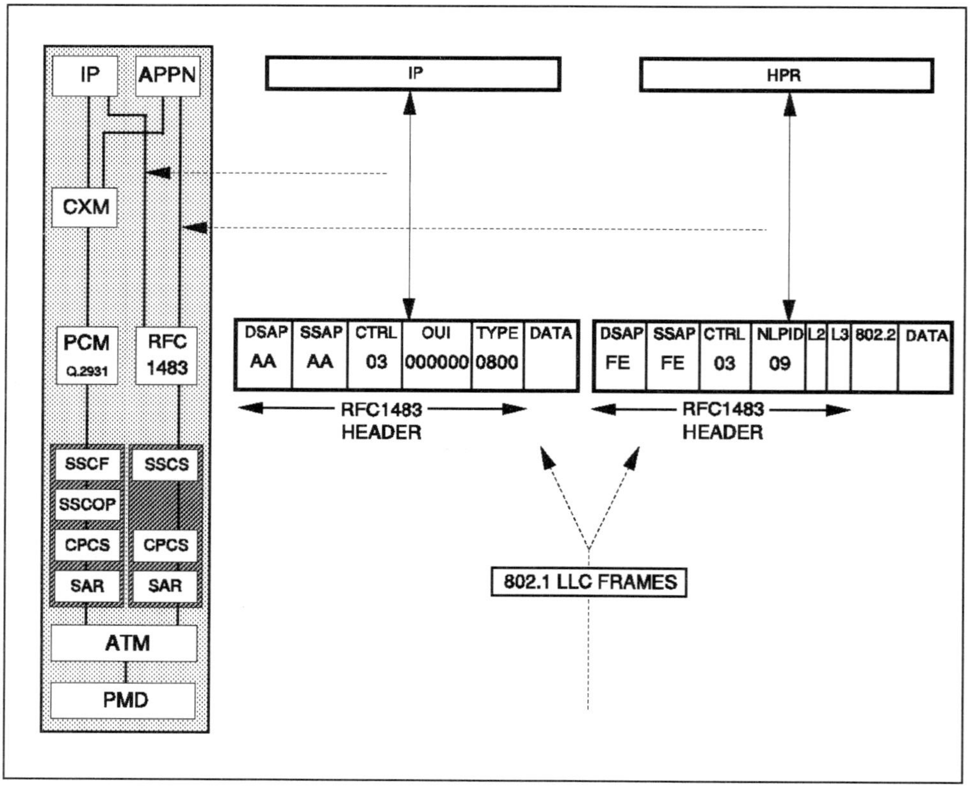

Figure 43. Multiprotocol Encapsulation on SVCs

RFC 1483 encapsulates packets of the various protocols within 802.2 LLC type 1 frames (see Figure 37 on page 96).

4.8 APPN Network Design Considerations

ATM technology is primarily associated with speed, although in many cases (some LAN applications excepted), data throughput problems are not effectively solved by adding more bandwidth. This is the reason that ATM provides traffic contracts and QOS. Traffic can be sent when bandwidth is not being used by other traffic, while high-priority traffic is given a guaranteed service. The native ATM DLC for APPN offers APPN an ideal way to use ATM QOS. Applications that use SNA's class of service (COS) can gain the benefits of ATM QOS without being changed.

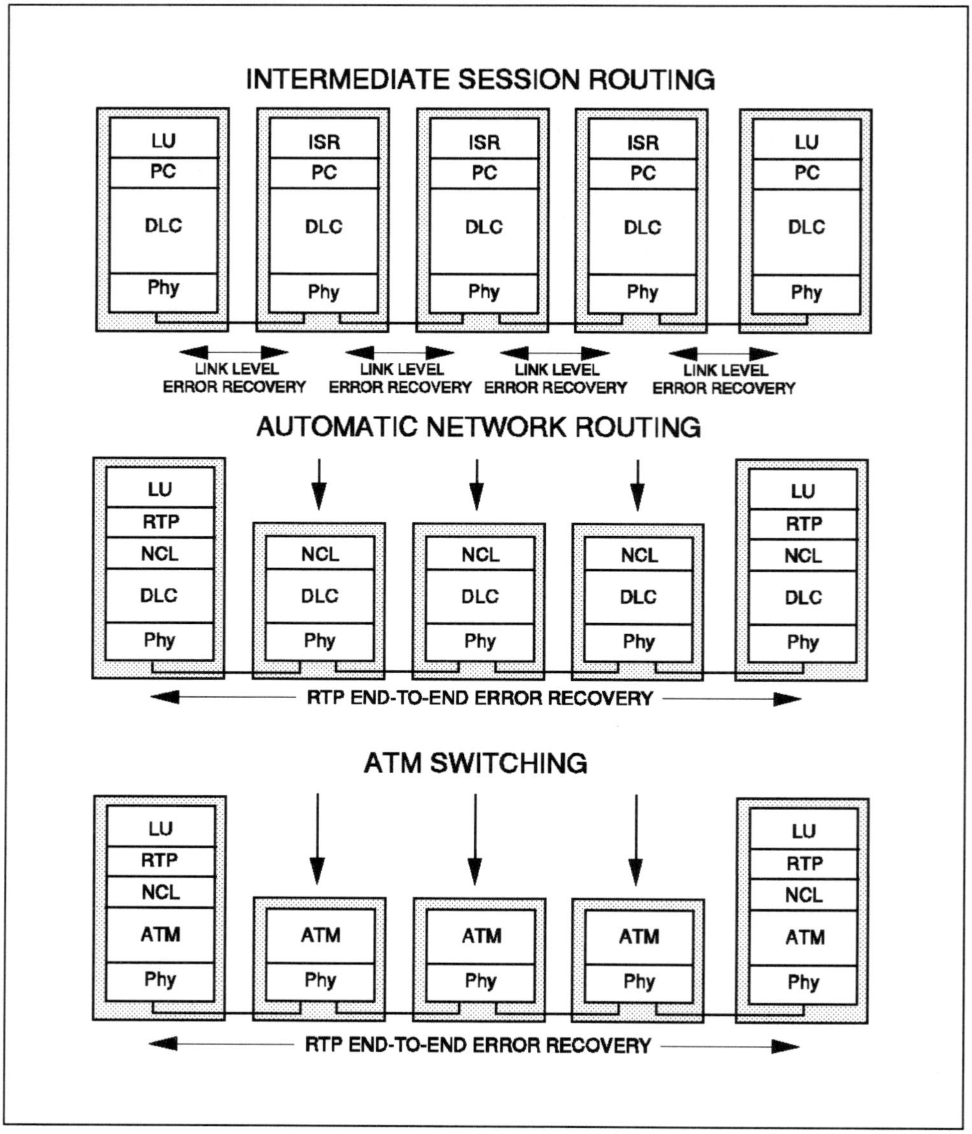

Figure 44. *APPN Routing Evolution*

APPN networks allow APPN nodes to route packets arriving on one link out on any other link. This intermediate session routing was further refined and made faster with the introduction of high performance routing (HPR). Session awareness was removed in intermediate nodes, with packets on the HPR connection being switched at the network control layer (NCL). An APPN native ATM DLC will replace ANR routing with ATM switching at the switching layer in intermediate nodes. Figure 44 shows this evolution. Depending upon whether ATM switching nodes will be APPN capable, or whether APPN nodes will be able to switch ATM cells, APPN networks will certainly need to be designed differently when using ATM. Switching at the ATM layer will effectively remove the intermediate nodes from APPN control. Where in earlier APPN networks each node along the session path had to be able to route APPN data, using ATM will remove the need for this prerequisite in intermediate nodes. Depending upon the network used between two APPN nodes, large wide area networks may increasingly start to look like connection networks, something that was restricted to LANs up until now. Certainly there is no need to have an APPN node along the session path if a direct virtual circuit is to be set up between the session endpoint nodes.

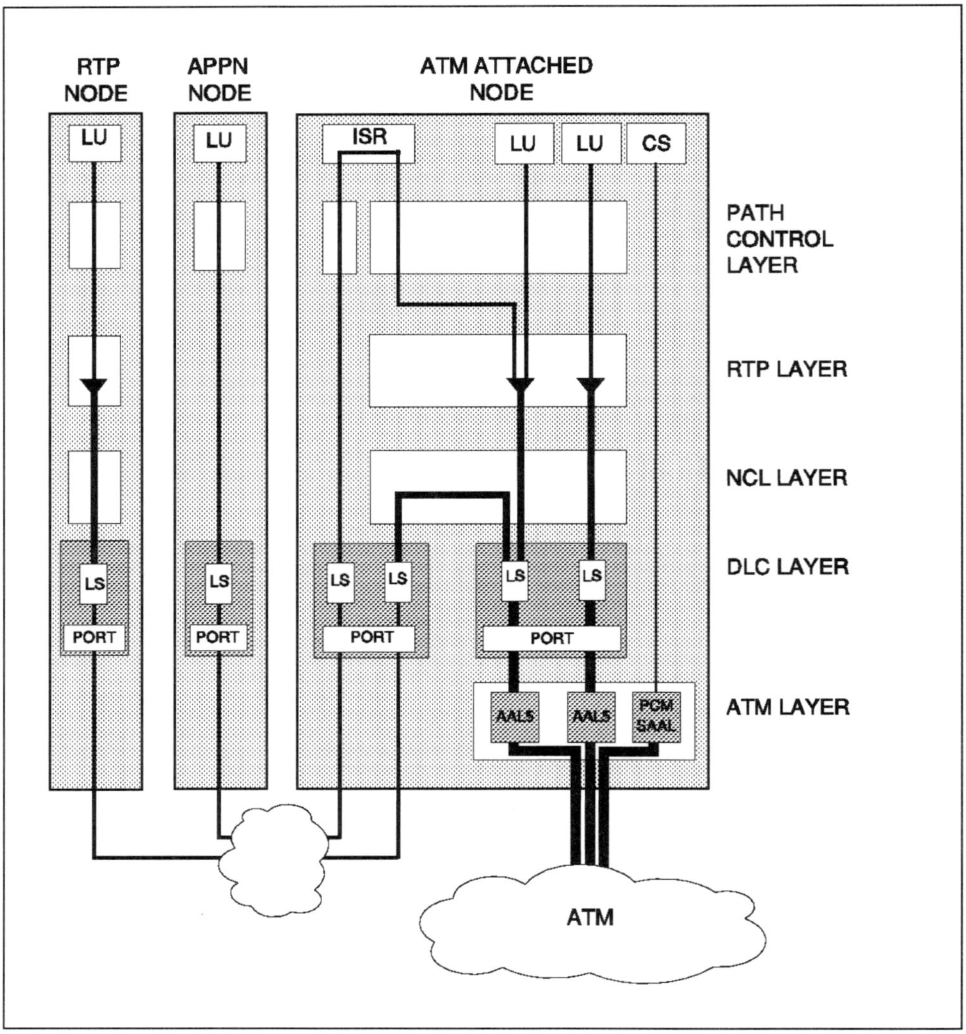

Figure 45. *APPN ATM Inter Node Flows*

Another factor in the design of future APPN networks will be the cost of ATM adapters (or access to the ATM WAN). In cases where a concentrator function is needed to allow a large number of APPN nodes (each with low bandwidth needs) to access an ATM WAN, adding a native ATM DLC to a network node at the edge of a WAN would be an ideal solution. Figure 45 shows such a concentrator function. The HPR capable node is able to route RTP and base APPN ISR connections over the ATM network.

A native ATM DLC also removes the need for another box between the APPN node and the ATM network, with all its implications for network management. APPN nodes with a native ATM DLC will have APPN and ATM network management built in, removing the problem of having to associate an APPN TG or link to an ATM VC on a separate management platform.

4.9 APPN High Performance Routing Compared to TCP/IP

The APPN extensions described in the previous sections explain how APPN nodes will be capable of creating direct VCCs to other APPN nodes in the network. An APPN node providing access to the ATM network will function like an edge device function group (EDFG) in MPOA. The APPN network nodes provide functions comparable to NHRP's next hop servers (NHS). The route returned allows a direct VCC to be established directly to an ATM-attached APPN node or to an egress point in the ATM network, that being an APPN node at the edge of the ATM network. Resource reservation, which resource reservation protocol (RSVP, see 8.3, "Resource Reservation Protocol (RSVP)" on page 189) will provide for IP networks, is already supplied by SNA's class of service to QOS mapping. Indeed it would seem that many of the extensions being proposed for IP networks to facilitate its use in ATM networks are either available already, or will soon be available in APPN networks.

In addition, HPR makes APPN equivalent to or better than TCP/IP in all areas:

- Path switching

  HPR has superior path switching that is COS-based and is performed end-to-end.

- Error recovery

  HPR's transport mechanism (RTP) performs the more efficient selective retransmission error recovery as opposed to TCP/IP's *go-back-n*.

- Data routing performance

  HPR's automatic network routing (ANR) is faster than TCP/IP's routing. If a packet is to be switched onto a link, ANR removes the first ANR label and passes the packet to DLC. An IP router must always decrement the *time to live* field in the IP header; this change means that the header checksum must be recomputed, then passed to DLC. IP header checksum recomputation is always a two-pass process, which is not optimal for high performance.

- Flow and congestion control

  HPR's adaptive rate based (ARB) flow/congestion control mechanism provides several features that make it superior to TCP/IP's:

- ARB will prevent network congestion rather than react to it.
- ARB produces steady rather than oscillating traffic patterns.
- ARB maximizes link utilization.

HPR is a proven network technology available in products today. The native ATM DLC architecture will allow HPR connections to share ATM SVCs with other protocols.

The ATM DLC architecture also supports the preservation of SNA's COS priorities across an ATM network by exploiting the multiple classes of service available directly from the ATM link layer. It does this by mapping multiple SNA streams of a particular SNA COS to a particular class of ATM VC set up for this purpose (for example, ABR), while mapping multiple SNA streams associated with a different, higher COS to a different higher class of ATM VC, such as VBR, all on the same ATM network interface. Existing SNA applications that utilize COS controls do not need to change in order to take advantage of this ATM functionality, as long as the SNA implementation has been designed to implement the native ATM DLC architecture.

Since today's standard IP networks do not support the concept of explicit priorities, the standard methods of mapping IP over ATM (such as Classical IP over ATM and LAN emulation) do not permit the kind of exploitation of ATM's multiple service classes described earlier for APPN.

There are several emerging protocols that are being defined for IP networks that deal with the concept of priority, all of which are still going through the IETF standardization process.

One of these protocols, RSVP, or Resource Reservation Protocol, together with a set of Integrated Services models, describe how priority is to be established for specific flows of IP packets, and how the priority is to be handled by each intermediate node in the network. The mapping of the Integrated Services models to specific link-layer technologies, such as ATM, is currently at an even earlier stage of standardization than the service models themselves.

RSVP will provide both a relative delay priority between data streams (Controlled Load Service), as well as specific delay guarantees for specific streams (Guaranteed Service), as requested by the application. An IP application, once it is modified to be able to request reservations via the RSVP API, can request simple priority handling for the packets associated with a specific data stream or can request the delay guarantees noted earlier. If the network cannot accept a requested bandwidth reservation for priority handling or for guaranteed service, the reservation request is rejected and the application notified. The forwarding path of intermediate network nodes (routers) must be specifically designed to support RSVP in order to classify the packets for priority queuing according to the port numbers and IP addresses carried in the IP and TCP/UDP headers, as well as to support the priority queuing at the network interfaces.

IPv6 is another emerging protocol that supports the concept of priority and which is currently going through an industry prototyping stage. It is not clear at this time when products based on IPv6 will begin to appear for general availability.

The IPv6 header has a priority field, as well as a flow label field. The flow label is intended to support the use of RSVP to create network bandwidth reservations as described earlier for current IP networks (IPv4). The IPv6 priority field could presumably be used to support the equivalent of SNA COS functionality, although the use of the priority field in IPv6 has yet to be fully defined. We are aware of no activity yet undertaken to map IPv6 priority fields to multiple ATM VCs with different ATM link-level classes of service, as we have done for APPN. In addition, IP applications will need to be modified in order to use new sockets interfaces for IPv6, which are still being defined.

Chapter 5. IP Support in ATM Networks

IP was the first network operating system that made use of ATM in a multivendor environment. Classical IP, or IETF RFC 1577, was the first standard available. Enhancements are made or proposed on all kinds of levels. This chapter gives an overview of work that is done or is under development and has IP as a base.

5.1 Classical IP over ATM (RFC 1577)

Since January 1993, the Internet Engineering Task Force (IETF) has had a formal recommendation on how to transport IP traffic over ATM. RFC 1577 (see 3 on page 229) describes the flows and mechanisms of Classical IP and ARP over ATM. RFC 1577 is the initial deployment of ATM within *classical* IP networks as a direct replacement for local area networks and for IP links that interconnect routers, either within or between administrative domains. The *classical* model refers to the treatment of the ATM host adapters as a networking interface to the IP protocol stack operating in a LAN-based paradigm.

Characteristics of the classical model are:

- The same maximum transmission unit (MTU) size is used for all VCs in a LIS.
- Default LLC/SNAP encapsulation of IP packets.
- End-to-end IP routing architecture stays the same.
- IP addresses are resolved to ATM addresses by use of an ATMARP service within the LIS; ATMARPs stay within the LIS. From a client's perspective, the ATMARP architecture stays faithful to the basic ARP model.
- One IP subnet is used for many hosts and routers. Each VC directly connects two IP members within the same LIS.

5.1.1 IP Subnetwork Configuration

In the LIS scenario, each separate administrative entity configures its hosts and routers within a closed logical IP subnetwork. Each LIS operates and communicates independently of other LISs on the same ATM network. Hosts connected to ATM communicate directly to hosts within the same LIS. Communication to hosts outside of the local LIS is provided via an IP router. This router is an ATM endpoint attached to the ATM network that is configured as a member of one or more LISs. This configuration may result in a number of disjoint LISs operating over the same ATM network. Hosts of differing OP subnets must communicate via an intermediate IP router even though it may be possible to open a direct VC between the two OP members over the ATM network.

The requirements for IP members (hosts, routers) operating in an ATM LIS configuration are:

- All members have the same IP network/subnet number and address mask.
- All members within a LIS are directly connected to the ATM network.
- All members outside of the LIS are accessed via a router.
- All members of a LIS must have a mechanism for resolving IP addresses to ATM addresses via ATMARP (based on RFC 826) and vice versa via InATMARP (based on RFC 1293) when using SVCs.
- All members of a LIS must have a mechanism for resolving VCs to IP addresses via InATMARP (based on RFC 1293) when using PVCs.
- All members within a LIS must be able to communicate via ATM with all other members in the same LIS; that is, the virtual connection topology underlying the intercommunication among the members is fully meshed.

The following list identifies a set of ATM specific parameters that must be implemented in each IP station connected to the ATM network:

- The ATM hardware address (atm$ha) is the ATM address of the individual IP station.
- The ATMARP request address (atm$arp-req) is the ATM address of an individual ATMARP server located within the LIS. In an SVC environment, ATMARP requests are sent to this address for the resolution of target protocol addresses to target ATM addresses. That server must have authoritative responsibility for resolving ATMARP requests of all IP members within the LIS.

  Note: If the LIS is operating with PVCs only, then this parameter may be set to null, and the IP station is not required to send ATMARP requests to the ATMARP server.

5.1.2 Permanent Virtual Connections

An IP station *must* have a mechanism (for example, manual configuration) for determining what PVCs it has and, in particular, which PVCs are being used with LLC/SNAP encapsulation.

All IP members supporting PVCs are required to use the Inverse ATM Address Resolution Protocol (InATMARP) (refer to RFC 1293) on those VCs using LLC/SNAP encapsulation. In a strict PVC environment, the receiver will infer the relevant VC from the VC on which the InATMARP request (InARP_REQUEST) or response (InARP_REPLY) was received. When the ATM source and/or target address is unknown, the corresponding ATM address length in the InATMARP packet *must* be set to zero (0) indicating a null length, otherwise the appropriate address field should be filled in and the corresponding length set appropriately.

5.1.3 Switched Virtual Connections

SVCs require support for ATMARP in the nonbroadcast, nonmulticast environment that ATM networks currently provide. To meet this need, a single ATMARP server must be located within the LIS. This server must have authoritative responsibility for resolving the ATMARP requests of all IP members within the LIS.

The server itself does not actively establish connections. It depends on the clients in the LIS to initiate the ATMARP registration procedure. An individual client connects to the ATMARP server using a point-to-point VC. The server, upon the completion of an ATM call/connection of a new VC specifying LLC/SNAP encapsulation, will transmit an InATMARP request to determine the IP address of the client. The InATMARP reply from the client contains the information necessary for the ATMARP server to build its ATMARP table cache. This information is used to generate replies to the ATMARP requests it receives.

The ATMARP server mechanism requires that each client be administratively configured with the ATM address of the ATMARP server (atm$arp-req) as defined earlier in this chapter. There is to be one and only one ATMARP server operational per logical IP subnet. It is recommended that the ATMARP server also be an IP station. This station must be administratively configured to operate and recognize itself as the ATMARP server for a LIS. The ATMARP server must be configured with an IP address for each logical IP subnet it is serving to support InATMARP requests.

5.1.4 Enhancing RFC 1577

In RFC 1577, it was not possible to have more than one ATMARP server within a LIS. In the future there will be an environment with either a single or multiple synchronized servers. To make an ATMARP server capable of supporting server-to-server neighbor synchronization protocol and operations, several extensions must be made to the single ATMARP server model. These changes are still under consideration and are not discussed further here.

The MTU size is becoming negotiable. The same maximum transmission unit (MTU) is the default for all VCs in a LIS. However, on a VC-by-VC point-to-point basis, the MTU size may be negotiated during connection startup using Path MTU Discovery to better suit the needs of the cooperating pair of IP members or the attributes of the communications path. The Path MTU Discovery mechanism is Internet Standard RFC 1191 and is an important mechanism for reducing IP fragmentation in the Internet. This mechanism is particularly important because the new subnet ATM uses a default MTU size significantly different from older subnet technologies, such as Ethernet and FDDI.

In order to ensure good performance through the Internet, and also to permit IP to take full advantage of the potentially larger IP datagram sizes supported by ATM, all router implementations that comply or conform with this specification must also implement the

IP Path MTU Discovery mechanism as defined in RFC 1191 and clarified by RFC 1435. Host implementations should implement the IP Path MTU Discovery mechanisms as defined in RFC 1191.

5.2 IP Address Resolution in ATM Networks

Address resolution protocols based on broadcasts to find the hardware or MAC address of the partner do not work in an ATM network. See 1.7.1.2, "ATM Networks" on page 29 for an explanation.

IP over ATM (RFC 1577) introduced the concept of a logical IP subnet (LIS). A LIS is a group of IP nodes that connect to a single ATM network and belong to the same IP subnet. Generally, communication between nodes in different IP subnets is only possible through an IP router, although a direct ATM VC between both partners would be possible. The disadvantages of this are clear:

- Reassembly of ATM cells into IP packets at the router so a routing decision can be made.
- No end-to-end QOS.
- Fast ATM switches connected by (relatively) slow routers.
- ATM cells may traverse the same ATM switch more than once.

In addition, inclusion of layer 3 routers means that a connectionless routing model is being overlaid on a connection-oriented transport layer.

To overcome the limitations previously explained, the Routing over Large Clouds (ROLC) working group of the IETF is working on protocols to overcome this limitation. The next hop resolution protocol (NHRP) is being proposed as a solution.

Nonbroadcast multiaccess NHRP extends the concept of a LIS. An NMBA network is one that allows multiple devices to be connected to the same network, but does not support broadcast between these nodes. ATM, frame relay and X.25 networks are examples of NBMA networks. For administrative or policy reasons, a physical NBMA network may be divided up into several *logical NBMA subnetworks*. A logical NBMA subnetwork is defined as a collection of hosts or routers that share *unfiltered subnetwork connectivity* over an NBMA subnetwork. Unfiltered subnetwork connectivity means that the nodes are not restricted from communicating with each other by things such as closed user groups or address screening.

NHRP has evolved from NARP, or RFC 1735, which was an experimental discussion. NHRP is a transport layer independent address resolution protocol that can be used in any NBMA network. The following section describes NHRP.

5.3 Next Hop Resolution Protocol (NHRP)

This section gives an explanation of NHRP as defined in the INTERNET DRAFT <draft-ietf-rolc-nhrp-07.txt>. This document can be found on http://ds.internic.net. This draft expired June 1996.

5.3.1 Introduction

The NBMA next hop resolution protocol (NHRP) allows a source station (a host or router), wishing to communicate over a nonbroadcast multiaccess (NBMA) subnetwork, to determine the internetworking layer addresses and NBMA addresses of suitable NBMA next hops toward a destination node. If the destination is connected to the NBMA subnetwork, then the NBMA next hop is the destination node itself. Otherwise, the NBMA next hop is the egress router from the NBMA subnetwork that is *nearest* to the destination node.

The main purpose of NHRP is to avoid routing hops in an NMBA network with multiple logical IP subnets. The advantage of this is clear, no layer 3 routing hops that need packet reassembly and end-to-end QOS through an ATM network.[9]

All NHRP packets, whether requester-to-server or server-to-server, are encapsulated in IP packets. This allows NHRP to be independent of the transport layer protocol being used.

An NBMA subnetwork will generally consist of multiple logical IP subnets (LISs). A LIS (see 3 on page 229 and 8 on page 229), has the following properties:

1. All members of a LIS have the same IP network/subnet number and address mask.
2. All members within a LIS are directly connected to the same NBMA subnetwork.
3. All members outside of the LIS are accessed via a router.

Address resolution (see 3 on page 229 and 8 on page 229) only resolves the next hop address if the destination node is a member of the same LIS as the source node; otherwise, the source station must forward packets to a router that is a member of multiple LISs. In multi-LIS configurations, hop-by-hop address resolution may not be sufficient to resolve the NBMA next hop toward the destination node, and IP packets may traverse the NBMA subnetwork more than once.

---

9 In the context of NHRP, the following terms are used:

- Internetwork layer: the media-independent layer (IP in the case of TCP/IP networks).
- Subnetwork layer: the media-dependent layer underlying the internetwork layer, including the NBMA technology (ATM, X.25, SMDS, etc.)

NHRP in its most basic form provides a simple internetworking layer to NBMA subnetwork layer address binding service. This may be sufficient for hosts that are directly connected to an NBMA subnetwork, allowing for straightforward implementations in NBMA nodes. NHRP also has the capability of determining the egress point from an NBMA subnetwork when the destination is not directly connected to the NBMA subnetwork, and the identity of the egress router is not learned by other methods (such as routing protocols). Optional extensions to NHRP provide additional robustness and diagnosability.

Address resolution techniques (see 3 on page 229 and 8 on page 229) may be in use when NHRP is deployed. ARP servers and services over NBMA subnetworks may be required to support hosts that are not capable of dealing with any model for communication other than the LIS model, and deployed hosts may not implement NHRP but may continue to support ARP variants (see 3 on page 229 and 8 on page 229). NHRP is intended to reduce or eliminate the extra router hops required by the LIS model and can be deployed in a noninterfering manner alongside existing ARP services.

The operation of NHRP to establish transit paths across NBMA subnetworks between two routers requires additional mechanisms to avoid stable routing loops (described in 5.3.7, "Stable Routing Loops" on page 127). Work is still underway to provide a solution to this problem.

5.3.2 NHRP Functional Components

Placed within the NBMA subnetwork are one or more entities that implement the NHRP protocol. Such nodes, which are capable of answering next hop resolution requests, are known as next hop servers (NHSs). Each NHS serves a set of destination hosts, which may or may not be directly connected to the NBMA subnetwork. NHSs cooperatively resolve the NBMA next hop within their logical NBMA subnetwork. In addition to NHRP, NHSs may participate in protocols used to disseminate routing information across (and beyond the boundaries of) the NBMA subnetwork and may support *classical ARP* service as well.

An NHS maintains a *next hop resolution* cache, which is a table of address mappings (internetwork layer address to NBMA subnetwork layer address). This table can be constructed from information learned from NHRP Register packets, extracted from next hop resolution requests/replies that traverse the NHS as they are forwarded, or through mechanisms outside the scope of this document (examples of such mechanisms include ARP (see 2 on page 229, 3 on page 229 and 8 on page 229) and preconfigured tables).

Whether or not a particular node within the NBMA subnetwork, which is making use of the NHRP protocol, needs to be able to act as an NHS is a local matter. For a node to avoid providing NHS functionality, there must be one or more NHSs within the NBMA subnetwork that are providing authoritative NBMA information on its behalf. If NHRP is to be able to resolve the NBMA address for nodes that lack NHS functionality, those serving NHSs must exist along all routed paths between next hop resolution requesters and the node that cannot answer next hop resolution requests.

A host or router that is not an NHRP server must be configured with the identity of the NHS that serves it (see 5.3.2.1, "Configuration").[10]

5.3.2.1 Configuration

Stations

To participate in NHRP, a node connected to an NBMA subnetwork should be configured with the NBMA address(es) of its NHS(s). Alternatively, it should be configured with a means of acquiring them (that is, the group address that members of an NHS group use for the purpose of address or next hop resolution). The NHS(s) will likely also represent the node's default or peer routers, so their NBMA addresses may be obtained from the node's existing configuration. If the node is attached to several subnetworks (including logical NBMA subnetworks), the node should also be configured to receive routing information from its NHS(s) and peer routers so that it can determine which internetwork layer networks are reachable through which subnetworks.

Next Hop Servers

An NHS is configured with knowledge of its own internetwork layer and NBMA addresses, a set of internetwork layer address prefixes that correspond to the internetwork layer addresses of the nodes it serves, and a logical NBMA subnetwork identifier. If a served node is attached to several subnetworks, the NHS may also need to be configured to advertise routing information to such nodes.

If an NHS acts as an egress router for nodes connected to other subnetworks than the NBMA subnetwork, the NHS must also be configured to exchange routing information between the NBMA subnetwork and the other subnetworks. In all cases, routing information is exchanged using conventional intradomain and/or interdomain routing protocols.

The NBMA addresses of the nodes served by the NHS may be learned via NHRP Register packets or manual configuration.

---

10 for NBMA subnetworks that offer group or multicast addressing features, it may be desirable to configure nodes with a group identity for NHSs, i.e., addressing information that would solicit a response from "all NHSs". The means whereby a group of NHSs divide responsibilities for next hop resolution are not described here.

5.3.3 Next Hop Resolution

In this section, we briefly describe how a source node (which potentially can be either a router or a host) uses NHRP to determine the NBMA next hop to a destination node.

NHRP supports two types of address resolution:

Authoritative Requests and Replies

An authoritative request is satisfied with an authoritative reply. Only the NHS that maintains the NBMA-to-internetwork layer mapping for a station can answer an authoritative request. An authoritative request will not be answered by an NHS that only has a cached entry for a station. If a communication attempt based on a nonauthoritative reply fails, a station can then choose to send an authoritative next hop resolution request.

Nonauthoritative Requests and Replies

An NHS that answers a nonauthoritative next hop resolution request with cached information issues a nonauthoritative next hop resolution reply.

If the NHS serves the station whose address resolution is being requested, it will reply to a nonauthoritative request with an authoritative reply.

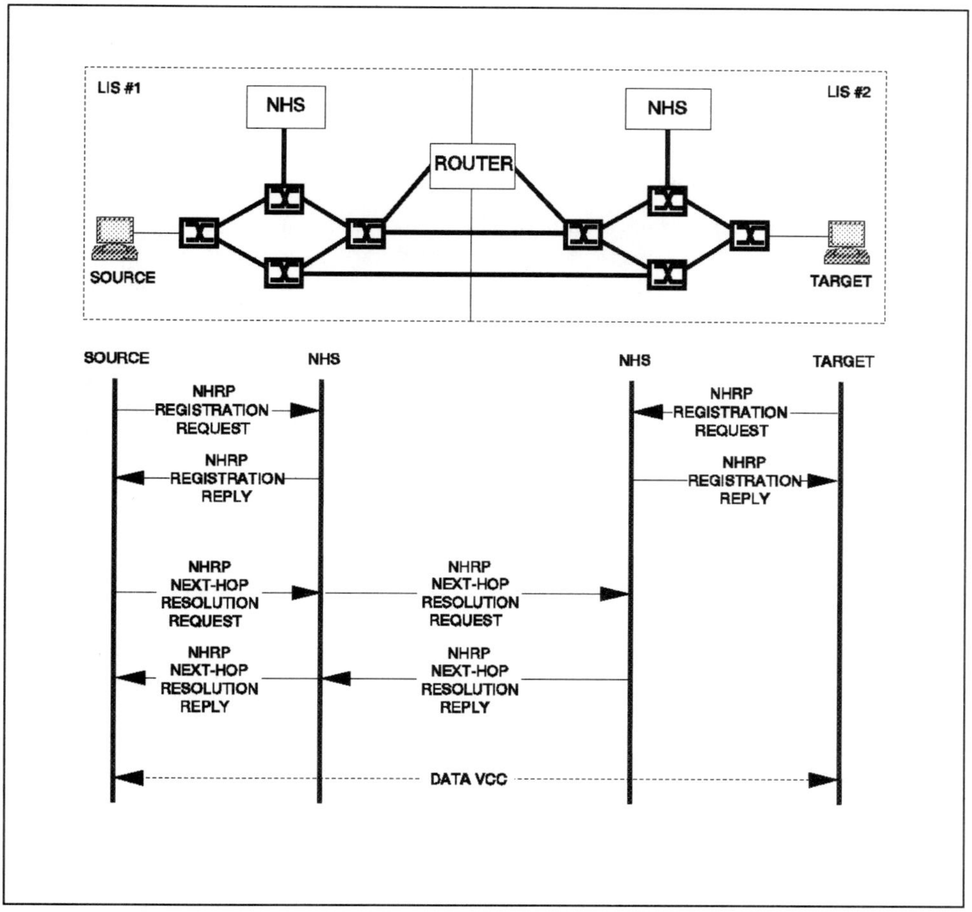

Figure 46. *NHRP Overview*

5.3.3.1 NHRP Flows

The sender first determines the next hop to the destination through normal routing processes (for a host, the next hop may simply be the default router; for routers, this is the next hop to the destination internetwork layer address). If the next hop is reachable through its NBMA interface, the sending node constructs a next hop resolution request packet containing the destination internetwork layer address as the (target) destination address, the sending nodes own internetwork layer address as the source address (next hop resolution request initiator), and senders NBMA addressing information. The sender also indicates whether it needs an authoritative or nonauthoritative reply. The sender emits the next hop resolution request packet towards the destination, using the NBMA address of the next routed hop. When the NHS receives a next hop resolution request, a check is made to see if it *serves* the desired destination (that is, the NHS checks to see if there is a next hop entry for the destination in its next hop resolution cache. If the NHS does not serve the destination, the NHS forwards the next hop resolution request to another NHS. Note that NHSs may only be one hop away from each other in order for forwarding of NHRP packets to be possible.

If this NHS serves the destination, the NHS resolves the destination's NBMA address and generates a positive next hop resolution reply on the destination's behalf. The next hop resolution reply packet contains the next hop internetwork layer address and the NBMA address for the destination and is sent back to the sender. (Note that if the destination is not on the NBMA subnetwork, the next hop internetwork layer address will be that of the egress router through which packets for the destination should be forwarded.)

An NHS receiving a next hop resolution reply may cache the NBMA next hop information contained therein. To a subsequent next hop resolution request, this NHS may respond with the cached, nonauthoritative NBMA next hop information or with cached negative information, if the NHS is allowed to do so.

The process of forwarding the next hop resolution request is repeated until the next hop resolution request is satisfied or an error occurs (for example, no NHS in the NBMA subnetwork can resolve the next hop resolution request). If it is determined that the destination's next hop cannot be resolved, a negative next hop resolution reply (NAK) is returned. This occurs when:

- No next hop resolution information is available for the destination from any NHS
- An NHS is unable to forward the next hop resolution request (for example, connectivity is lost)

NHRP optionally provides a mechanism to send a next hop resolution reply that contains aggregated NBMA next hop information. Suppose that a router is the NBMA next hop from a sending node to the destination. Suppose further that the router is an egress router for all nodes sharing an internetwork layer address prefix with the destination. When a next hop resolution reply is generated in response to a next hop resolution request, the responder may augment the internetwork layer address of the destination with a prefix length. A subsequent (nonauthoritative) next hop resolution request for some destination that shares an internetwork layer address prefix (for the number of bits specified in the prefix length) with the destination may be satisfied with this cached information. See 5.3.5, "Cache Management Issues" on page 124 regarding caching issues.

To dynamically detect subnetwork-layer filtering in NBMA subnetworks (for example, X.25 closed user group facility or SMDS address screens), as well as to provide loop detection and diagnostic capabilities, a *route record* may be included in NHRP packets. The route record extensions contain the internetwork (and subnetwork layer) addresses of all intermediate NHSs between source and destination (in the forward direction) and between destination and source (in the reverse direction). When a source node is unable to communicate with the responder (for example, an attempt to open an SVC fails), it may attempt to do so successively with other subnetwork layer addresses in the route record until it succeeds (if authentication policy permits such action). This approach can find a suitable egress point in the presence of subnetwork-layer filtering (which may be source/destination sensitive, for instance, without necessarily creating separate logical NBMA subnetworks) or subnetwork-layer congestion (especially in connection-oriented media).

The result of a next hop resolution request depends on how routing is configured among the NHSs of an NBMA subnetwork. If the destination node is directly connected to the NBMA subnetwork and the routed path to it lies entirely within the NBMA subnetwork, the next hop resolution replies always return the NBMA address of the destination node itself rather than the NBMA address of some egress router. On the other hand, if the routed path exits the NBMA subnetwork, NHRP will be unable to resolve the NBMA address of the destination, but rather will return the address of the egress router. For destinations outside the NBMA subnetwork, egress routers and routers in the other subnetworks should exchange routing information so that the optimal egress router may be found.

In addition to NHSs, an NBMA node could also be associated with one or more regular routers that could act as connectionless servers for the node. The station could then choose to resolve the NBMA next hop or just send the packets to one of its connectionless servers. The latter option may be desirable if communication with the destination is short-lived and/or doesn't require much network resources. The connectionless servers could, of course, be physically integrated in the NHSs by augmenting them with internetwork layer switching functionality.

5.3.3.2 Routing NHRP Requests and Replies

NHRP registration requests, NHRP registration replies, NHRP purge requests, NHRP purge replies, and NHRP error indications follow the routed path from sender to receiver in the same fashion that next hop resolution requests and next hop resolution replies do. That is, requests and indications follow the routed path from Source Protocol Address (which is the address of the node initiating the communication) to the Destination Protocol Address. Replies, on the other hand, follow the routed path from the Destination Protocol Address back to the Source Protocol Address except for the case when a next hop resolution reply is sent directly back to the requester via a direct VCC to reduce response time.

Next hop resolution requests and next hop resolution replies *must not* cross the borders of a logical NBMA subnetwork (an explicit NBMA subnetwork identifier may be included as an extension in the next hop resolution request). Thus, the internetwork layer traffic out of and into a logical NBMA subnetwork always traverses an internetwork layer router at its border. Internetwork layer filtering can then be implemented at these border routers.

Note: A next hop resolution reply can be returned directly to the next hop resolution request initiator (that is, without traversing the list of NHSs that forwarded the next hop resolution request) if all of the following criteria are satisfied:

1. Direct communication is available via datagram transfer (for example, SMDS) or the NHS has an existing virtual circuit connection to the next hop resolution request initiator or is permitted to open one.
2. The next hop resolution request initiator has not included the NHRP reverse NHS record extension.
3. The authentication policy in force permits direct communication between the NHS and the next hop resolution request initiator.

The purpose of allowing an NHS to send a next hop resolution reply directly is to reduce response time. A consequence of allowing a direct next hop resolution reply is that NHSs that would under normal circumstances be traversed by the next hop resolution reply would not cache next hop information contained therein.

5.3.3.3 Triggering a Resolution Request

The most likely trigger for a next hop resolution request is that a data packet addressed to the destination is to be emitted from the sending node (either because the sender is a host or is a transit router). Address resolution could also be triggered by other means (for example, a routing protocol update packet).

5.3.3.4 Waiting for a Resolution Reply

If the next hop resolution request is triggered by a data packet, the sender may choose to dispose of the data packet while awaiting a next hop resolution reply in one of the following ways:

1. Drop the packet
2. Retain the packet until the next hop resolution reply arrives and a more optimal path is available
3. Forward the packet along the routed path toward the destination

   NHRP does not address, however, any possible packet misordering that may be caused.

The choice of which of the previous functions to perform is a local policy matter, though option (3) is the recommended default, since it may allow data to flow to the destination while the NBMA address is being resolved.

5.3.4 Deployment

Next hop resolution requests traverse one or more hops within an NBMA subnetwork before reaching the node that is expected to generate a response. Each node, including the source station, chooses a neighboring NHS to which it will forward the next hop resolution request. The NHS selection procedure typically involves performing a routing decision based upon the network layer destination address of the next hop resolution request. Ignoring error situations, the next hop resolution request eventually arrives at a node that is to generate a next hop resolution reply. This responding node either serves the destination or is the destination itself, if both NHRP client and server functionality are coresident in the same node. The responding station generates a next hop resolution reply using the source address from within the NHRP packet to determine where the next hop resolution reply should be sent.

The next hop resolution request packet is carried at the NBMA layer, with a destination NBMA address set to that of the locally determined NHS. If the addressed NHS does not serve the destination address specified in the next hop resolution request, the next hop resolution request packet is routed at the network layer based upon the next hop resolution requester′s destination address and forwarded to the neighboring NHS determined by the routing decision. Alternately, the NHS may use static configuration information in order to determine to which neighboring NHSs to forward the next hop resolution request packet. Each NHS/router examines the next hop resolution request packet on its way toward the destination, optionally modifying it on the way (such as updating the forward record extension), and continues to forward it until it reaches the NHS that serves the destination network layer address.

In order to forward NHRP packets to a neighboring NHS, NHRP clients must nominally be configured with the NBMA address of at least one NHS. In practice, a client′s default router should also be its NHS. A client may be able to derive the NBMA address of its NHS from the configuration that was already required for the client to be able to communicate with its next hop router.

Forwarding of NHRP packets within an NBMA subnetwork requires a contiguous deployment of NHRP capable nodes. During migration to NHRP, it cannot be expected that all nodes within the NBMA subnetwork are NHRP capable. NHRP traffic that would otherwise need to be forwarded through such nodes can be expected to be dropped due to the NHRP packet being unrecognized. In this case, NHRP will be unable to establish any transit paths whose discovery requires the traversal of the non-NHRP speaking nodes. If the client has tried and failed to acquire a cut-through route, the client should use the network layer routed path as a default.

The path taken by next hop resolution requests will normally be the same as the path taken by data packets that are routed at the network layer to the desired destination. (The paths may be different in situations where NHSs have been statically configured to forward traffic by other means. For example, a next hop resolution request may be forwarded to a group multicast address.)

NHSs should acquire knowledge about destinations other NHSs serve as a direct consequence of participating in intradomain and interdomain routing protocol exchange. In this case, the NHS serving a particular destination must lie along the routed path to that destination. In practice, this means that all egress routers must double as NHSs serving the destinations beyond them, and that hosts on the NBMA subnetwork are served by routers that double as NHSs.

NHSs (and end nodes) may alternately be statically configured with the NBMA addresses of their neighbors, the identities of the destinations that each of them serves and, optionally, a logical NBMA subnetwork identifier. Such static configurations may be necessary in cases where NHSs do not contain network layer routing protocol implementations.

If the NBMA subnetwork offers a link layer group addressing or multicast feature, the client (node) may be configured with a group address assigned to the group of next hop servers. The client might then submit next hop resolution requests to the group address, eliciting a response from one or more NHSs, depending on the response strategy selected.

NHSs may also be deployed with the group or multicast address of their peers, and an NHS might use this as a means of forwarding next hop resolution requests it cannot satisfy to its peers. This might elicit a response (to the NHS) from one or more NHSs, depending on the response strategy. The NHS would then forward the next hop resolution reply to the next hop resolution request originator. The purpose of using group addressing or a similar multicast mechanism in this scenario would be to eliminate the need to preconfigure each NHS in a logical NBMA subnetwork with both the individual identities of other NHSs as well as the destinations they serve. It reduces the number of NHSs that might be traversed to process a next hop resolution request (in those configurations where NHSs either respond or forward via the multicast, only two NHSs would be traversed) and allows the NHS that serves the next hop resolution request originator to cache next hop information associated with the next hop resolution reply.

5.3.5 Cache Management Issues

The management of NHRP caches in the source node, the NHS serving the destination, and any intermediate NHSs is dependent on a number of factors.

5.3.5.1 Caching Requirements

Source Stations: Source nodes *must* cache all received next hop resolution replies that they are actively using. They also must cache incomplete entries (that is, those for which a next hop resolution request has been sent but which a next hop resolution reply has not been received). This is necessary in order to preserve the Request ID for retries and provides the state necessary to avoid triggering next hop resolution requests for every data packet sent to the destination.

Source nodes *must* purge expired information from their caches. Source nodes *must* purge the appropriate cached information upon receipt of an NHRP purge request packet.

When a node has a coresident client and NHS, the station may reply to next hop resolution requests with information that the node cached as a result of the station making its own next hop resolution requests and receiving its own next hop resolution replies, as long as the node follows the rules for Transit NHSs.

Serving NHSs

The NHS serving the destination (the one which responds authoritatively to next hop resolution requests) *should* cache information about all next hop resolution requests to which it has responded, if the information in the next hop resolution reply has the possibility of changing during its lifetime (so that an NHRP purge request packet can be sent). The NBMA information provided by the source node in the next hop resolution request may be cached for the duration of its holding time. This information is considered to be stable, since it identifies a node directly attached to the NBMA subnetwork. An example of unstable information is NBMA information derived from a routing table, where that routing table information has not been guaranteed to be stable through administrative means.

Transit NHSs

A Transit NHS (lying along the NHRP path between the source node and the responding NHS) may cache information contained in next hop resolution request packets that it forwards. A Transit NHS may cache information contained in next hop resolution reply packets that it forwards only if that next hop resolution reply has the stable (B) bit set. It *must* discard any cached information whose holding time has expired. It may return cached information in response to nonauthoritative next hop resolution requests only.

5.3.5.2 Dynamics of Cached Information

NBMA-Connected Destinations

NHRP's most basic function is that of simple NBMA address resolution of nodes directly attached to the NBMA subnetwork. These mappings are typically very static, and appropriately chosen holding times will minimize problems in the event that the NBMA address of a node must be changed. Stale information will cause a loss of connectivity, which may be used to trigger an authoritative next hop resolution request and bypass the old data. In the worst case, connectivity will fail until the cache entry times out.

This applies equally to information marked in next hop resolution replies as being *stable* (via the B bit).

This also applies equally well to source nodes that are routers, as well as those which are hosts.

Note that the information carried in the next hop resolution request packet is always considered stable because it represents a node that is directly connected to the NBMA subnetwork.

Destinations Outside the NBMA Network

If the source of a next hop resolution request is a host, the destination is not directly attached to the NBMA subnetwork, and the route to that destination is not considered to be stable, then the destination mapping may be very dynamic (except in the case of a subnetwork where each destination is only singularly homed to the NBMA subnetwork). As such, the cached information may very likely become stale. The consequence of stale information in this case will be a suboptimal path (unless the internetwork has partitioned or some other routing failure has occurred).

Strategies for maintaining NHRP cache information in the presence of dynamic routing changes are still being resolved.

5.3.6 The NHRP Domino Effect

One could easily imagine a situation where a router, acting as an ingress node to the NBMA subnetwork, receives a data packet, such that this packet triggers a next hop resolution request. If the router forwards this data packet without waiting for an NHRP transit path to be established, then when the next router along the path receives the packet, the next router may do exactly the same (that is, originate its own next hop resolution request, as well as forward the packet). In fact, such a data packet may trigger next hop resolution request generation at every router along the path through an NBMA subnetwork. We refer to this phenomena as the NHRP *domino* effect. The NHRP domino effect is clearly undesirable. At best it may result in excessive NHRP traffic. At worst it may result in an excessive number of virtual circuits being established unnecessarily. Therefore, it is important to take certain measures to avoid or suppress this behavior. NHRP implementations for NHSs *must* provide a mechanism to address this problem. It is recommended that implementations provide one or more of the following solutions:

- Possibly the most straightforward solution for suppressing the domino effect would be to require transit routers to be preconfigured not to originate next hop resolution requests for data traffic that is simply being forwarded (not originated). In this case,the routers avoid the domino effect through an administrative policy.
- A second possible solution would be to require that when a router forwards a next hop resolution request, the router instantiates a (short-lived) state. This state consists

of the route that was used to forward the next hop resolution request. If the router receives a data packet, and the packet triggers a next hop resolution request generation by the router, the router checks whether the route to forward the next hop resolution request was recently used to forward some other next hop resolution request. If so, then the router suppresses generation of the new next hop resolution request (but still forwards the data packet). This solution also requires that a node should attempt to originate a next hop resolution request before the data packet that triggered the next hop resolution request. Otherwise, unnecessary next hop resolution requests may still be generated.

- A third possible strategy would be to configure a router in such a way that next hop resolution request generation by the router would be driven only by the traffic the router receives over its non-NBMA interfaces (interfaces that are not attached to an NBMA subnetwork). Traffic received by the router over its NBMA-attached interfaces would not trigger NHRP next hop resolution requests. Just as in the first case, such a router avoids the NHRP domino effect through administrative means.
- Finally, rate limiting of next hop resolution requests may help to avoid the NHRP domino effect. Intermediate routers that would otherwise generate unnecessary next hop resolution requests may instead suppress such next hop resolution requests due to the measured next hop resolution request rate exceeding a certain threshold. Of course, such rate limiting would have to be very aggressive in order to completely avoid the domino effect. Further work is needed to analyze this solution.

5.3.7 Stable Routing Loops

One problem still under discussion as NHRP is being standardized is how to avoid so-called *stable routing loops*. Stable routing loops are not limited to NHRP networks. With NHRP, stable routing loops can occur when a backdoor route is available between two routers that are outside of the NBMA. Figure 47 on page 128 shows two routers; these are both connected to the NBMA, and they also have a backdoor connection.

A stable routing loop can occur in the configuration shown in Figure 47 on page 128 when the following happens:

1. Router #1 starts forwarding packets from the source to the target over the NBMA network.
2. The path between router #2 and the target breaks.
3. Router #2 now looks for an alternate path to the target.
4. Routing tables show that router #1 has a path to the subnet where the target resides; this route is the backdoor path between the two routers.
5. Router #2 forwards the packets to router #1 who dutifully forwards them into the NBMA network, causing the packets to loop.

Such stable routing loops can occur because NHRP does not know anything about paths outside of the NBMA network. NHRP cannot know that a path behind router #2 is broken; it also knows nothing about the backdoor path between the two routers. This can lead to a loop because a cut-through route between two routers is only used for data transport and not for router-router traffic. IP routing protocols assume that router-router updates are sent across all paths that data is sent across.

There are various solutions to the stable routing loop problem:

- NHRP should only resolve addresses for targets inside the NBMA network. In this case, paths across the NBMA to an egress router would only be via normal routing protocols. NHRP would not supply a cut-through route for such paths.
- NARP also contained a method to avoid stable routing loops. NARP only supplied address resolution for end systems, not routers, that are outside of the NBMA network.

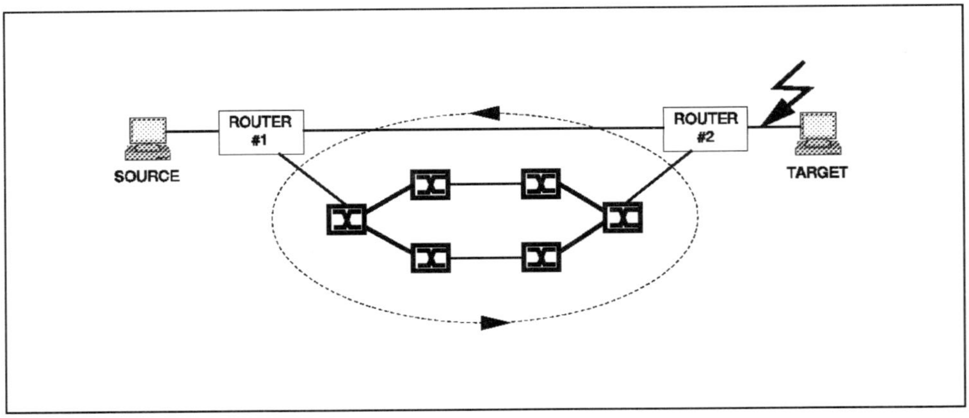

Figure 47. *Stable Routing Loops*

5.3.8 NHRP in ATM Networks

The process of sending a next hop resolution request, waiting as that request is passed from next hop server to next hop server, receiving the resolution reply, and then establishing a direct data VCC to the target can take a great deal of time. Its obvious that this overhead is not acceptable for shorter connections, such as name resolution requests and pings, whereas FTP sessions will make better use of the throughput and QOS that an ATM direct VCC provides. To make the use of NHRP feasible, end stations and routers will need to be able to differentiate between types of IP traffic; and so, decide when a next hop resolution request should be issued and when the packets should only be forwarded to the next layer 3 router.

NHRP will be used as the shortcut route resolution protocol for MPOA.

5.4 IP Multicasting in ATM Networks

At the transport layer, certain LAN types directly support broadcast or multicast operations. Ethernet, for example, uses the low-order bit of the high-order byte to distinguish unicast addresses from multicast addresses.

At the networking layer, IP does not directly support multicast operations, although IP nodes that implement a joint protocol that uses IGMP could receive data in a multicast group. RFC 1577 allows mapping a multicast IP address to an ATM address, but does not deal with functions, such as registration to a multicast group, or how IP multicast groups would be mapped to the underlying ATM layer to form a multicast group.

5.4.1 ATM Multicast Approaches

Two methods of implementing multicast mechanisms over ATM networks are detailed in the following sections. Figure 48 on page 130 shows both methods.

5.4.1.1 VC Meshes

The most fundamental approach to intracluster multicasting is the multicast VC mesh. Each source establishes its own independent point-to-multipoint VC (a single multicast tree) to the set of leaf nodes (destinations) that it has been told are members of the group to which it wishes to send packets.

Interfaces that are both senders and group members (leaf nodes) to a given group will originate one point-to-multipoint VC and terminate one VC for every other active sender to the group. Multipoint-to-multipoint communication is achieved when multiple senders establish their own data paths to the same set of leaf (receiving) nodes. This criss-crossing of VCs across the ATM network gives rise to the name *VC mesh*.

5.4.1.2 Multicast Servers

An alternative model has each source establish a VC to an intermediate node, the multicast server (MCS). The multicast server itself establishes and manages a point-to-multipoint VC out to the actual desired destinations.

The MCS reassembles AAL SDUs arriving on all the incoming VCs, and then queues them for transmission on its single, outgoing point-to-multipoint VC. Reassembly of incoming AAL SDUs is required at the multicast server as AAL5 does not support cell level multiplexing of different AAL SDUs on a single outgoing VC.

The leaf nodes of the multicast server's point-to-multipoint VC must be established prior to packet transmission, and the multicast server requires an external mechanism to identify them.

5.4.1.3 VC Meshes versus Multicast Servers

Ultimately the choice depends on the relative trade-offs a system administrator must make between throughput, latency, congestion, and resource consumption, as shown in Figure 48. The characteristics of the applications generating the multicast traffic would be a major influence.

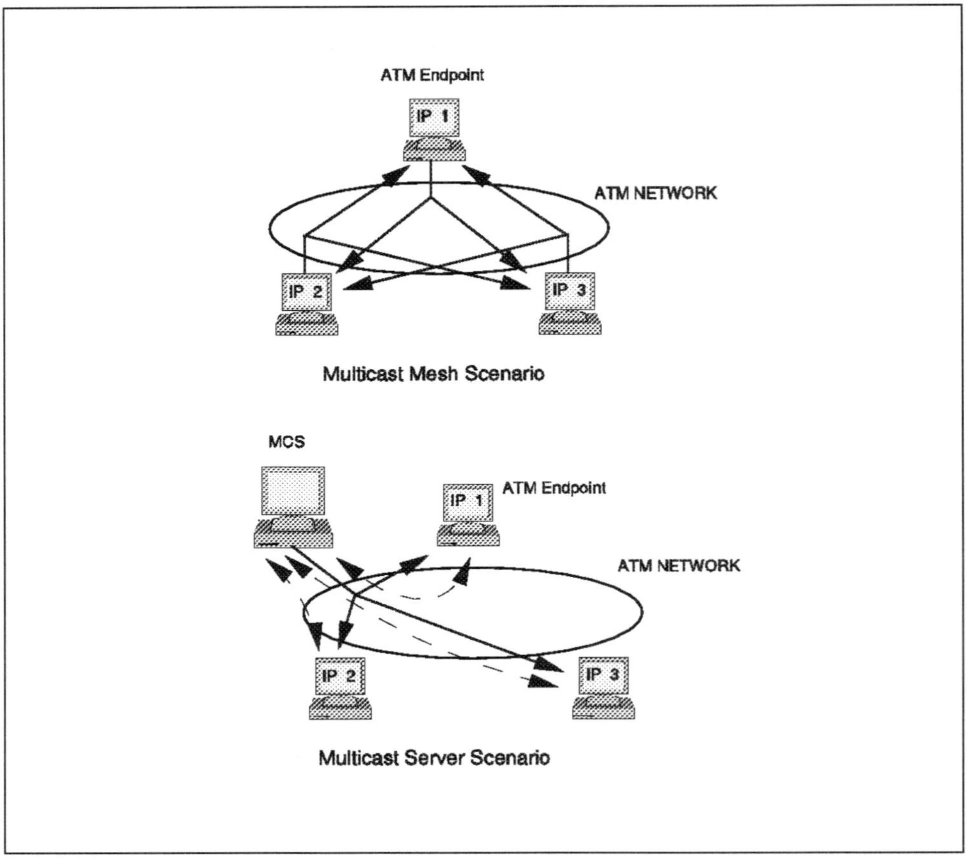

Figure 48. *ATM Multicast Options*

If we focused on the data path, we might prefer the VC mesh because it lacks the obvious single congestion point of a multicast server. Throughput is likely to be higher and end-to-end latency lower, because the mesh lacks the intermediate AAL SDU reassembly that must occur in MCSs. The underlying ATM signalling system also has greater opportunity to ensure optimal branching points at ATM switches along the multicast trees originating on each source. However, consumption of connection resources will be higher; every group member's ATM interface must terminate a VC per sender. On the contrary, with a multicast server, only two VCs (one out and one in) are required, independent of the number of senders.

If we focus on the signalling load, then MCSs have the advantage when faced with dynamic sets of receivers. Every time the membership of a multicast group changes (a leaf node needs to be added or dropped), only a single point-to-multipoint VC needs to be modified when using an MCS. This generates a single signalling event across the MCS's UNI. However, when membership change occurs in a VC mesh, signalling events occur at the UNIs of every traffic source; the transient signalling load scales with the number of sources. This has obvious ramifications for the convergence time taken for a group's connectivity to stabilize after change.

Finally, multicast servers introduce a *reflected packet problem*. Sources that are also group members will get copies of their own packets straight back from the multicast server. The MPOA solution must ensure sufficient information is carried within each PDU to enable identification and removal of these reflected packets (the problem is addressed in LANE by the use of a LECID in the packet encapsulation).

5.4.2 Multicast Address Resolution Server (MARS)

The IETF draft (see 11 on page 230) introduces a *Multicast Address Resolution Server* (MARS), which evolved from the ATMARP server introduced in RFC 1577. Where the ARP server keeps a table of address mappings with {IP, ATM} for all IP endpoints in a LIS, the MARS keeps extended tables with {LAYER 3, ATM.1, ATM.2, ... ATM.n} mappings. Although MARS is a mechanism to support the multicast needs of layer 3 protocols, in general, the first implementations will most likely be seen for IPv4.

As the LLC/SNAP code points for MARS are different than those for ARP, MARS and ARP server functionality may be implemented within a common entity and share a client server VC, if the implementer so chooses.

Clusters

In practice, a cluster is the set of endpoints that chooses to use the same MARS to register their memberships and receive their updates from.

By implication of this definition, traffic between interfaces belonging to different clusters passes through an intercluster device. In the IP world, an intercluster device would be an IP multicast router with logical interfaces into each cluster.

The term *cluster member* is used to refer to an endpoint that is currently using a MARS for multicast support. Thus the potential scope of a cluster may be the entire membership of a LIS, while the actual scope of a cluster depends on which endpoints are actually cluster members at any given time.

MARS Client

A MARS client or endpoint is best thought of as being a layer between the layer 3 protocols link-layer interface and the UNI 3.0/3.1 interface. It can exist in a host or router.

The client establishes a bidirectional VC to the MARS; this is used to send queries and receive replies. It is suggested that this VC be dismantled if not used for a period of time.

The other signalling path is a cluster control VC to which the client is added as a leaf node when it registers with the MARS.

5.4.3 MARS Operation

MARS supports both multicast server and multipoint mesh operation. The choice of which to use may be made on a per-group basis and is transparent to the endpoints.

5.4.3.1 MARS Multicast Servers

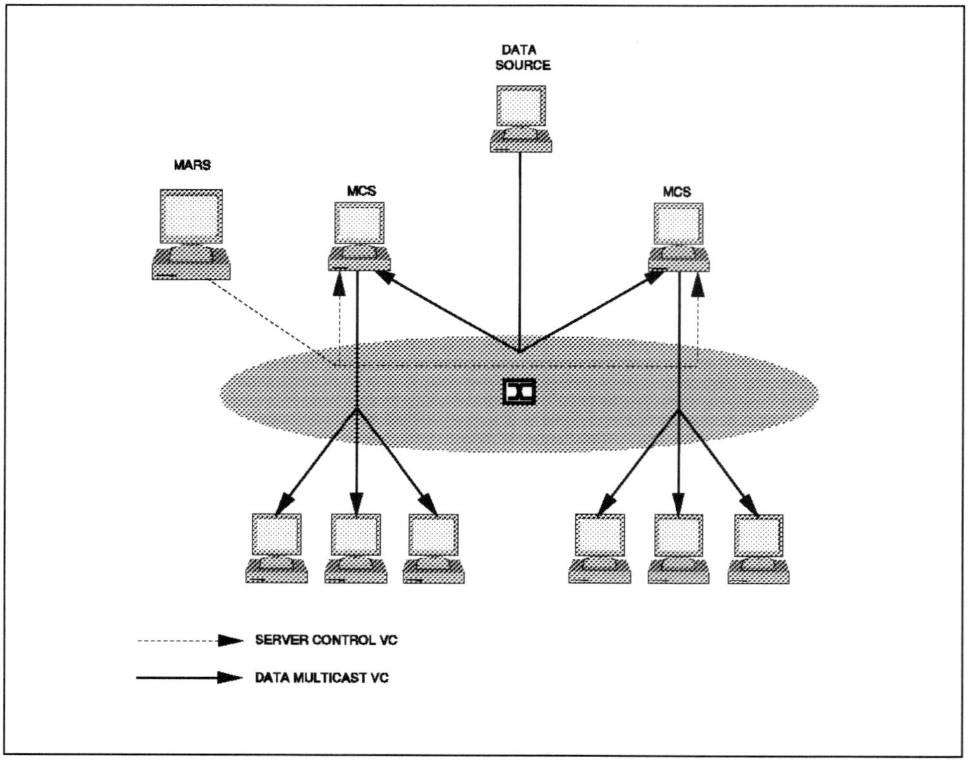

Figure 49. Multicast Server

Registration

A multicast server that wishes to serve a multicast group registers with the MARS. The registration message is used to construct a server map for each multicast address; these are the addresses returned when a multicast address is requested for that group. The MARS adds the server to its server control VC. This VC is used to inform the servers when group membership changes occur.

Resolution

The MARS returns a message that contains a *map* of ATM addresses of one or more multicast servers serving the group. The requesting node uses this map to set up connections, point-to-point in the case of a single server or point-to-multipoint if multiple servers are returned, to the multicast servers and then transmits the multicast packets on those connections. The multicast servers have point-to-multipoint VCs to the nodes that they serve, and incoming packets are re-sent down the multipoint VC to the targets.

The MARS does not take part in the actual multicasting of the data packets.

5.4.3.2 MARS Multipoint Meshes

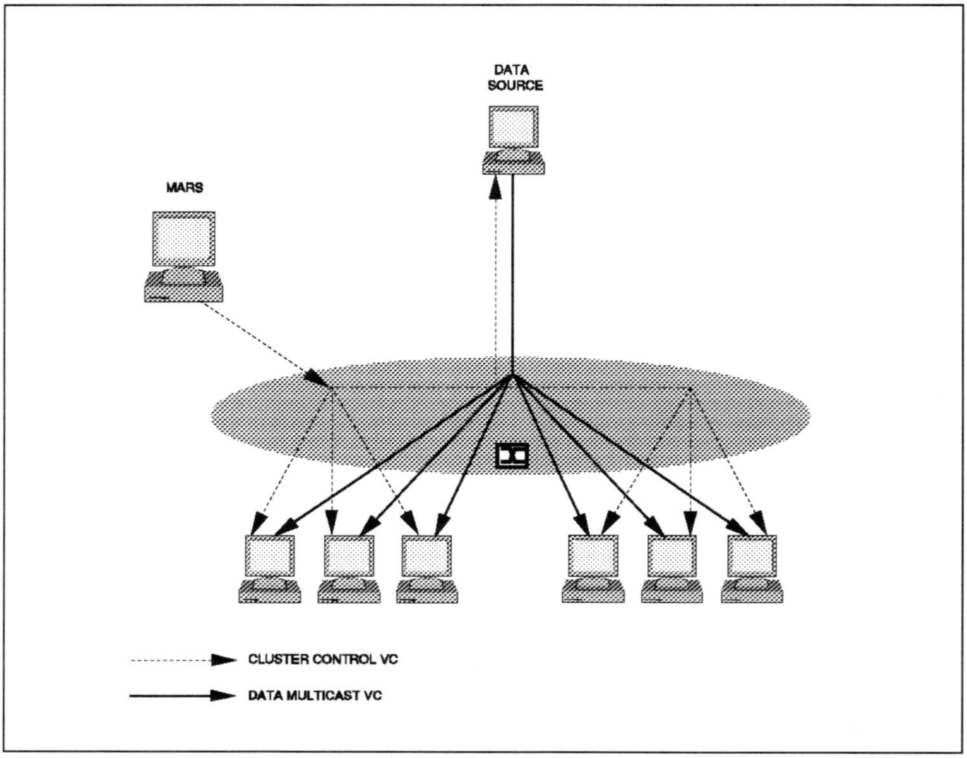

Figure 50. Single Multicast Mesh

Registration

Any node wishing to join or transmit data to a multicast group must first register with the MARS. MARS adds this endpoint as a leaf on its cluster control VC. This VC is used by the MARS to inform all endpoints when a node joins or leaves a group so that the endpoints can add or remove them from their point-to-multipoint VCs.

Resolution

An endpoint that registers with the MARS gets a list of the ATM addresses of nodes that are registered as members of that multicast group. The endpoint then builds a point-to-multipoint connection to all those nodes and sends the multicast data on that connection. This packets go directly to each target via the VC.

The MARS does not take part in the actual multicasting of the data packets.

Chapter 6. PNNI Phase 1 and Integrated PNNI

In the layered routing model, layer-3 packets (for example, IP and IPX) are forwarded by routers on a hop-by-hop basis based on the contents of the destination network address. A dynamic routing protocol such as OSPF is used to distribute network topology and reachability information amongst the routers in the network. In ATM networks a similar mechanism is required to forward SVC requests through a network of switches based on the called or destination ATM address. Likewise a dynamic routing protocol can be used to distribute ATM network topology and reachability information amongst the switches in the network. This will enable ATM switches to compute an accurate path through the network to the destination.

Due to the unique scaling and QOS features of ATM as well as the connection-oriented nature of SVCs, the ATM routing protocol must address several important requirements:

- Must be stable and reflect the current topology and capacity of the network.
- Must scale from a network of several switches to one containing hundreds and perhaps thousands.
- Must support efficient and loop-free routing of SVC requests over a path that will meet the QOS objectives of the SVC.
- Must support a heterogeneous mix of ATM switches.
- Must be extensible. That is it should be simple to add and enhance new function as requirements arise.

The ATM forum is currently working on the Private Network-to-Network Interface Specification Version 1.0 (PNNI Phase I) which is designed to meet the above requirements.

The ATM Forum is also examining ways to leverage and exploit the PNNI Phase I protocol as a means to support layer-3 internetworking over ATM. Two other specifications are under development and will be discussed:

PNNI Augmented Routing (PAR)
: PAR allows IP routers directly connected to ATM switches to participate in PNNI and use the information gained from PNNI for establishing SVCs with other routers attached to the same ATM network.

Integrated PNNI (I-PNNI)
I-PNNI is an extension to PNNI Phase I in which a network of routers and switches run a single routing protocol that supports both SVC and packet routing. This single protocol is I-PNNI.

6.1 PNNI Overview

The PNNI protocol is intended for use between ATM switches in a private ATM network. The abbreviation PNNI stands for either Private Network Node Interface or Private Network-to-Network Interface and to some degree reflects the duality of its capability based on its recursive behavior as described below. The PNNI Phase I specification defines two distinct protocols:

PNNI Routing Protocol
This protocol is responsible for distributing topology information between switches in an ATM network. This information is used to compute a path through the network for the SVC that will satisfy the requested QOS. A hierarchy mechanism ensures that this protocol scales well for large ATM networks.

PNNI Signalling Protocol
This protocol defines the signalling flows used to establish point-to-point and multipoint connections across the ATM network. The protocol is based on standard ATM UNI signalling (3.1 and 4.0). PNNI signalling employs source routing and a crankback mechanism to route SVC requests around failed network components at call setup.

6.2 PNNI Design Concepts

To adapt to the stringent and complex demands of an ATM network, the developers of PNNI Phase I incorporated and in fact borrowed a number of techniques and algorithms that are present in many of today's routing protocols such as OSPF and BGP. PNNI is a topology state routing protocol. This is similar to common link-state routing protocols such as OSPF. It differs slightly in that the status of a node (in this case, an ATM switch) is advertised in addition to the status of the links. Otherwise it enjoys all of the important benefits of link-state routing. These advantages become quite apparent when placed in the context of a large ATM network with the requirement to successfully route many dozens if not hundreds and thousands of SVC requests per second. It converges rapidly when topology change occurs. It provides ATM switches with a "picture" of the current network topology. This enables a loop free path that will meet the requested QOS objectives of the SVC request to be computed. There is less network traffic overhead when compared with standard distance vector protocols.

PNNI networks can generate a hierarchical topology that reflects the topology and addressing of the network of ATM switches. This is similar to the 2-level hierarchy that OSPF can employ. Switches that share a common addressing scheme are grouped into an area or what PNNI refers to as a *peer group*. Members of a peer group will exchange information with each other about the topology of the peer group. A single ATM switch, called the peer group leader (PGL) then summarizes this information and exchanges it with other PGLs that represent other peer groups of switches in the next higher layer peer group. PNNI Phase I supports up to 104 levels of hierarchy. Event the largest networks will never need more than three or four. The topology illustrated in Figure 51 depicts two levels of hierarchy.

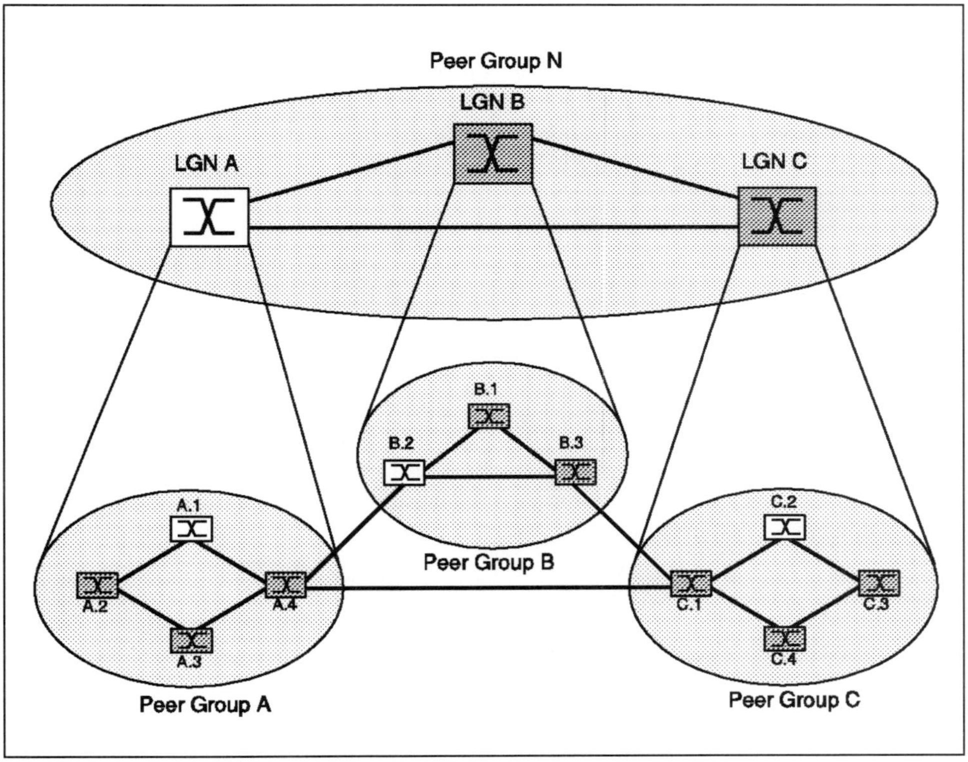

Figure 51. *PNNI Routing Hierarchy*

The three peer groups at the bottom of the figure represent a topology of real ATM switches connected by physical links. The switches in peer group A, for example, will exchange topology and resource information with the other switches in the peer group. Switch A.1 is elected the PGL and will summarize the information about peer group A. In the next higher-level peer group, peer group N, the PGL for peer group A, switch A.1 will exchange the summarized information with the other nodes in peer group N. The other PGLs representing peer groups B and C will do likewise. Switch A.1 will then advertise the summarized information it has gleaned from the other members of peer group N into its own lower level or child peer group, peer group A.

Thus, each switch in a peer group will have complete information about the topology of the peer group it is part of and partial or summarized information about the outside or external peer groups. Hierarchy enables a network to scale by reducing the amount of information a node is required to maintain. It contains the amount of real topology information that is transmitted on the network to a local area or peer group. The information on the network is further reduced by the process of topology aggregation so that a collection of real switches can appear as a single node to other peer groups.

Another mechanism that the PNNI developers borrowed was source routing. Simply put, a switch that receives an SVC request from a device over a UNI connection will compute and generate the entire path through the network. It will designate which switches the SVC request should pass through. This is called a designated transit list (DTL). If the SVC request is destined for a switch in another peer group it will specify all the external peer groups the SVC should travel through and direct it to a border switch in an adjacent peer group. It will be up to the entry or border switch of the adjacent or intermediate peer group to generate a DTL for its peer group. Source routing enables loop free paths to be computed and off-load route processing from the intermediate switches along the path. It also provides a certain degree of latitude and flexibility to enable switch vendors to implement their choice of route generation procedures.

In addition to topology state routing, hierarchy and source routing, PNNI Phase I was designed to be extensible. This enables new function and extensions to be incorporated into PNNI while preserving a backwards compatibility. This is accomplished by encoding PNNI advertised information in type/length/value (TLV) fields. PNNI has also defined procedures that enable a node to ignore information it does not need to know about.

6.3 PNNI Routing

The PNNI routing protocol is responsible for distributing topology information between switches in an ATM network. The following section explains how PNNI routing protocol works.

6.3.1 Addressing

The fundamental purpose of PNNI is to compute a route from a source to a destination based on a called ATM address. The called ATM address is an information element contained in the SETUP message that is sent over the UNI from the device to a switch (Section 5.4.5.11 of the ATM UNI 3.1 specification). Presumably a switch running PNNI Phase I will have in its topology database an entry that will match a portion or prefix of the 20-byte ATM address that is contained in the SETUP message. The switch will then be able to compute a path through the network to the destination switch.

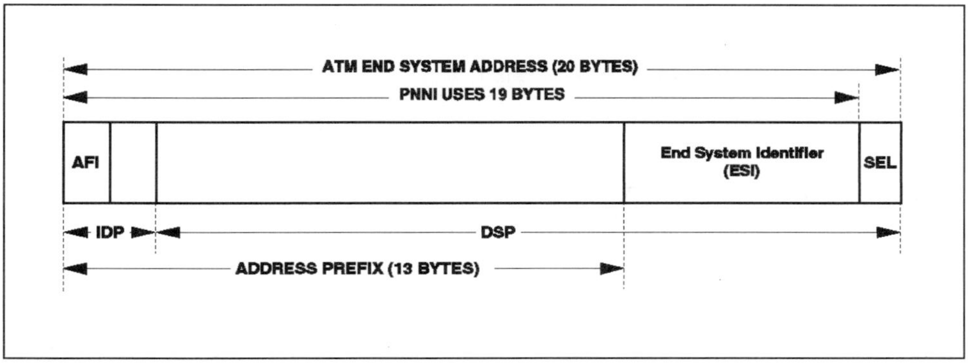

Figure 52. *ATM End-System Address*

To best understand PNNI routing, it would be helpful to review ATM addressing from PNNI's perspective. Addressing and identification of components of the PNNI routing hierarchy are based on the use of ATM end system addresses. An ATM end system address is 20 bytes long and is shown in Figure 52. PNNI routing works off of the first 19 bytes of this address or some prefix of this address. The 20th byte is the selector field which only has local significance to the end station and is ignored by PNNI routing. A prefix is the first "p" bits of an ATM address with "p" being a value between 0 and 152. Generally speaking PNNI will advertise reachability to ATM end systems using an address prefix rather than advertise each unique ATM end-system address. This is analogous to routers advertising reachability to IP subnets rather than to each individual IP host.

Nodes in a peer group have the same prefix address bits in common. This is illustrated in Figure 53 on page 140.

- At the highest level illustrated, the LGNs that make up the high-order LGN have their left *x* high-order bits the same.
- At the next lower level, the three LGNs shown have their left *x+y* high-order bits the same.
- At the lowest level illustrated, the LGNs have their left *x+y+z* high-order bits the same. (At this level, they are real physical switches.)

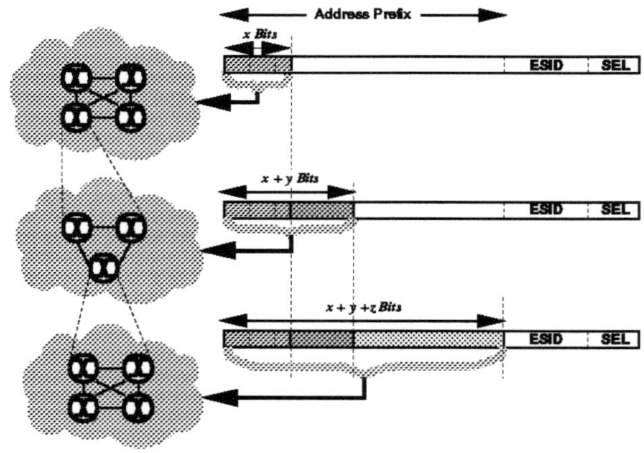

***Figure 53.** Addressing Hierarchy*

Two identifiers are used in PNNI to define the hierarchy and a nodes placement in the hierarchy. The first is the peer group identifier. This is a 14-byte value and is illustrated in Figure 54. The first byte is a level indicator which defines which of the next 104 left-most bits are shared by switches in the peer group. Peer group identifiers must be prefixes of ATM addresses.

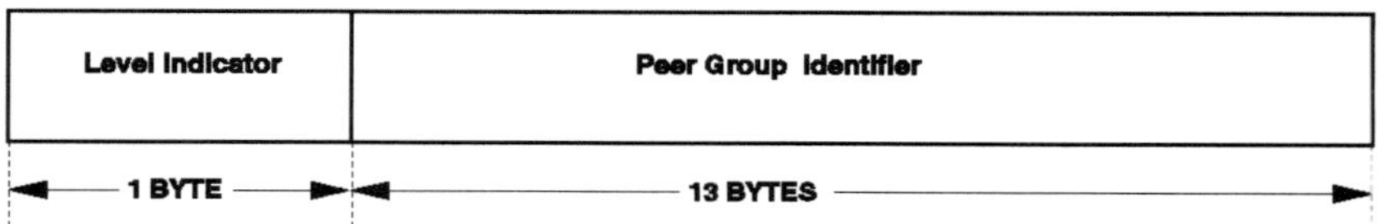

***Figure 54.** Peer Group Identifier*

A peer group is identified by its peer group identifier. Peer group IDs are specified at configuration time. Neighboring nodes exchange peer group IDs in hello packets. If they have the same peer group ID then they belong to the same peer group. If the exchanged peer group IDs are different, then the nodes belong to different peer groups.

The node identifier is 22 bytes in length and consists of a 1-byte level indicator and a unique 21-byte value. The node identifier is unique for each PNNI node in the routing domain. A PNNI node that advertises topology information in PNNI topology state packets will include the node identifier and the peer group identifier to indicate the originator of the information and the scope (on which level of the hierarchy it is directed to).

6.3.2 PNNI Information Exchange

A PNNI node will advertise its own direct knowledge of the ATM network. The scope of this advertisement is the peer group. The information is encoded in TLVs called PNNI Topology State Elements (PTSE). Multiple PTSEs can be carried in a single PNNI Topology State Packet (PTSP). The PTSP is the packet used to send topology information to a neighbor node in the peer group. Each switch advertises the following:

Nodal Information
: This includes the switch's ATM address, peer group identifier, leadership priority and other aspects about the switch itself.

Topology State Information
: This covers outbound link and switch resources.

Reachability
: ATM addresses and ATM address prefixes that the switch has learned about or is configured with.

It was mentioned that PNNI is a topology state protocol. This means that logical nodes will advertise link state and nodal state parameters. A link state parameter describes the characteristics of a specific link and a nodal state parameter describes the characteristics of a node. Together these can form topology state parameters that are advertised by PNNI nodes within their own peer group.

Topology state parameters are either metrics or attributes. A topology state metric is a parameter whose values must be combined for all links and nodes in the SVC request path to determine if the path is acceptable. A topology state attribute is a parameter that is considered individually to determine if a path is acceptable for an SVC request. Topology state attributes can be further subdivided into two categories: performance-related and policy-related. Performance-related attributes measure the performance of a particular link or node. Policy-related attributes provide a measure of conformance level to a specific policy by a node or link in the topology.

Table 8.1 shows the topology state parameters supported by PNNI and an explanation of each follows.

Table 1. PNNI Topology State Parameters		
Metrics	**Performance/Resource Attributes**	**Policy Attributes**
Cell Delay Variation	Cell Loss Ratio for CLP=0	Restricted Transit Flag
Maximum Cell Transfer Delay	Maximum Cell Rate	
Administrative Weight	Available Cell Rate	
	Cell Rate Margin	
	Variance Factor	
	Branching Flag	

Cell Delay Variation (CDV)

Expected CDV along the path relevant for CBR and VBR-rt traffic.

Administrative Weight (AW)

Link or nodal state parameter set by administrator to indicate preference.

Cell Loss Ratio (CLR)

Describes the expected CLR at a node or link for CLP=0 traffic.

Maximum Cell Rate (MCR)

Describes the maximum link or node capacity.

Available Cell Rate (ACR)

Measure of effective available bandwidth of the advertiser.

Cell Rate Margin (CRM)

Measure of difference between effective bandwidth allocation and the allocation for sustainable cell rate (SCR).

Variance Factor (VF)

Relative measure of the square of the CRM normalized by the variance of the aggregate cell rate.

Branching Flag

Used to indicate if a node can branch point-to-multipoint traffic.

Restricted Transit Flag
Nodal state parameter that indicates whether a node supports transit traffic or not.

6.3.3 PNNI Routing Hierarchy

The process of generating a PNNI routing hierarchy is an automatic procedure that defines how nodes will interact with each other. It begins at the lowest level in the hierarchy and is based on the information that is exchanged between switches. The same process is performed at each level of the hierarchy. To illustrate this process we return to the topology shown in Figure 51 on page 137 and begin with peer group A.

- Switches in peer group A exchange hello packets with their neighbor switches over a special reserved VCC (VPI=0, VCI=18) called the routing control channel. Hello packets contain a node's ATM end system address, node ID, and its port ID for the link. In this way the hello protocol makes the neighboring nodes known to each other.
- Membership in the peer group is determined based on addressing. Those with a matching peer group identifier are common peer group members.
- Topology information in the form of PTSEs is reliably flooded in the peer group over the routing control channel. PTSEs are the smallest collection of PNNI routing information that is flooded as a unit among all logical nodes within a peer group. A node's topology database consists of a collection of all PTSEs received, which represent that node's present view of the PNNI routing domain. The topology database provides all the information required to compute a route from the given node to any address reachable through that routing domain
- A peer group leader (PGL) is elected based on the leadership priority configured in the switch. The PGL represents the peer group as a logical group node in the next higher-level peer group. In Figure 51 on page 137, switch A.1 is the PGL for peer group A. A logical group node (LGN) is an abstract representation of a lower-level peer group for the purposes of representing that peer group in the next higher-layer peer group. LGN A represents peer group A in the next higher-level peer group, peer group N. Because PNNI is recursive, LGN A behaves just like it was a switch in a peer group which in this case is peer group N. It is also the responsibility for the PGL to advertise PTSEs that it has collected in higher-level peer groups. This enables the switches in peer group A to have at least a partial picture of the entire network.
- Identify uplink and build horizontal links between LGNs. An uplink is a connection to an adjacent peer group. This is discovered when border switches exchange hellos and determine that they are not in the same peer group. From the perspective of a switch in peer group A; an uplink is a connection to an LGN in a higher-level peer group. A horizontal link is a logical connection between LGNs in the next higher-level peer group. It is in actuality an SVC between PGLs. So the horizontal link that connects LGN A and LGN B in peer group N is an SVC between switches

A.1 and B.2. It functions as a routing control channel so that nodes in peer group N can exchange topology information.

- The same process of exchanging hellos and flooding PTSEs is performed in peer group N.

6.3.4 Generic Connection Admission Control (GCAC)

CAC is the function performed by ATM switches that determines whether a connection request can be accepted or not. This is performed by every switch in the SVC request path. But CAC is not standardized and it is up to the individual switch to decide if a connection request and its associated QOS can be supported.

PNNI uses information stored in the originating node's topology database, along with the connection's traffic characteristics and QOS requirements, to compute a path. But again CAC is a local switch process that the originating node cannot realistically keep track of. Therefore PNNI invokes a Generic Connection Admission Control (GCAC) procedure during the path selection process which provides the originating node with an estimate on whether each switch's local CAC process will accept the connection.

6.4 PNNI Signalling

PNNI signalling is based on a subset of UNI 4.0 signalling. It does not support some of the UNI Signalling 4.0 signalling features such as proxy signalling, leaf initiated join capability or user-to-user supplementary, but does support new capabilities such as specific QOS parameters, ATM anycast addressing and scoping and ABR. In addition, PNNI signalling differs from UNI 4.0 signalling in that it is symmetric. This makes sense because this is a switch-to-switch signalling protocol rather than a end-user-to-switch protocol.

PNNI signalling utilizes information gathered by PNNI routing. Specifically, it uses the route calculations derived from reachability, connectivity, and resource information dynamically maintained by PNNI routing. These routes are calculated as needed from the node's view of the current topology.

The unique capabilities that PNNI signalling defines are designated transit lists and crankback and alternate routing. DTLs are used to carry hierarchically-complete source routes. Crankback and alternate routing allows for an SVC request to be rerouted around a failed component as it makes its way to the destination switch. Additionally associated signalling is used for PNNI operation over virtual path connections and soft permanent VPCs/VCCs are supported.

6.4.1 Designated Transit Lists

PNNI defines routes using designated transit lists (DTLs). A DTL is a complete path across a peer group consisting of a sequence of node IDs and optionally port IDs of every switch that the SVC request is to traverse. DTLs are provided by the source node (switch that received the SVC request over the UNI) or an entry border node to a peer group. A hierarchically complete source route represents a route across a PNNI routing domain. This is expressed as a stack of source routes. Each source route in the stack represents the complete path across the specific level of the hierarchy.

In Figure 55 on page 146, User_A connected to switch A.2 wishes to establish an SVC with User_C. User_A generates an SVC request (SETUP message) and sends it over the UNI to switch A.2. Switch A.2 generates a stack of DTLs. The first one establishes a source route across peer group A. The second pertains to the next higher-level in hierarchy and is the complete path across peer group N. When the SVC request reaches peer group B, the top DTL in the stack is removed and switch B.2 generates a new DTL that defines a source route across peer group B. The same procedure occurs when the SVC request reaches peer group C.

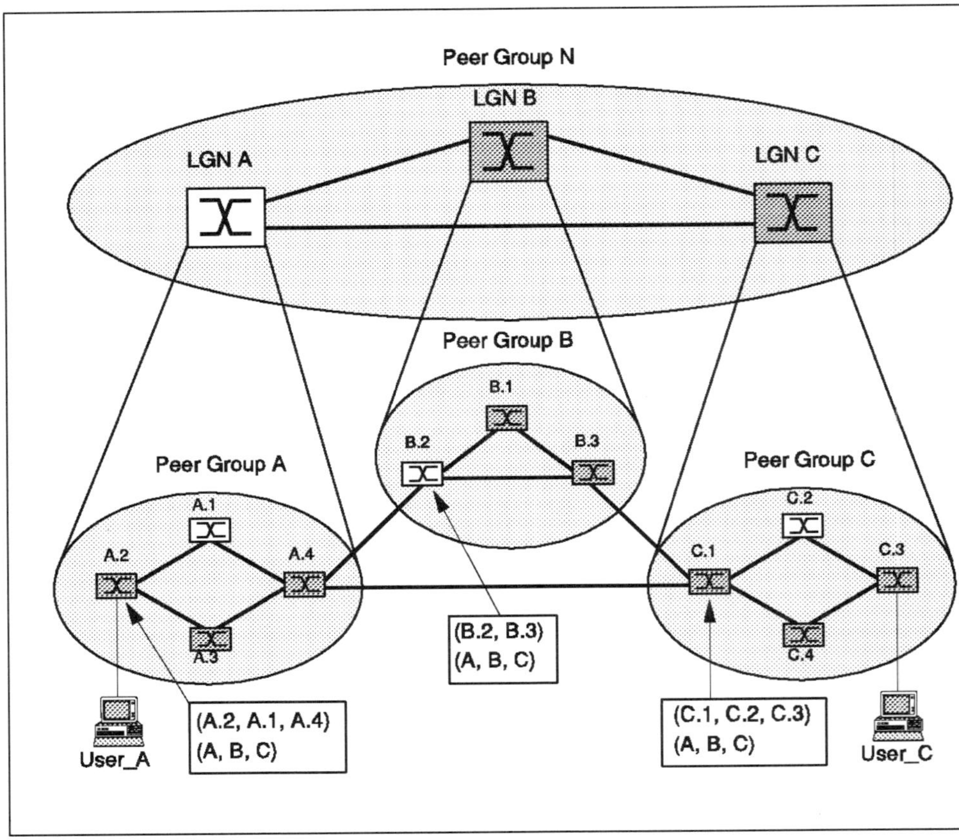

Figure 55. PNNI Designated Transit Lists

DTLs are created by the originating switch and/or entry border nodes. DTLs are encoded as information elements contained in the SETUP or ADD PARTY messages.

6.4.2 Crankback and Alternate Routing

When creating a DTL, a node uses the currently available information about resources and connectivity. This information may be inaccurate for a number of reasons. These reasons include hierarchical aggregation (aggregation leads to inaccuracy) and changes in resource availability due to additional calls that have been placed since the information was produced. Therefore an SVC request being processed according to the DTL may be blocked along its specified route. Crankback and alternate routing is a mechanism used for adapting to this situation short of clearing the call back to the source. When an SVC request cannot be processed according to the DTL, it is cranked back to the creator of that DTL with an indication of the problem. This node may choose an alternate path over which to progress the call or may further crankback the call. An alternate path must obey all received higher-level DTLs, and must avoid the blocked node(s) or link(s).

Crankback with alternate routing is illustrated in Figure 56. The entry border node in peer group B generates a DTL that will route the SVC request directly to switch B.3. But something happens in the port at switch B.3 that blocks the SVC request. So switch B.3 cranks it back to the DTL originator, switch B.2, who can either crank it further back to switch A.2 or generate a new DTL which bypasses the failed component. Switch B.2 chooses the latter and a new DTL is generated that routes the SVC request around the failed component. Notice that the new DTL conforms with the route specified in the higher-level DTL.

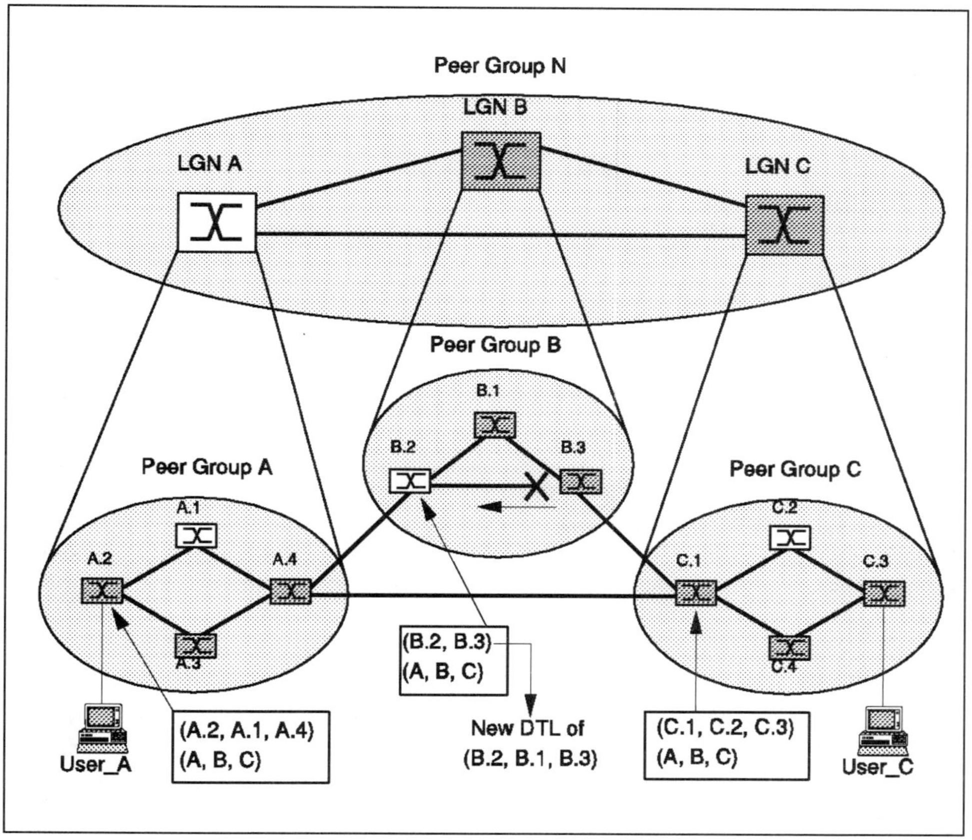

***Figure 56.** Crankback with Alternate Routing*

6.5 PNNI Phase I Summary of Features

PNNI Phase I has the following characteristics:

- Supports all UNI 3.1 and some UNI Signalling 4.0 functions
- Scales to very large networks
- Supports hierarchical routing
- Supports QOS
- Supports multiple routing metrics and attributes
- Uses source routing
- Reroutes around failed components at connection setup
- Supports soft PVC/PVP (administratively define connection setup and PNNI used to establish connection)
- Supports anycast
- Supports extensibility

6.6 PNNI Augmented Routing

Routers or route servers attached to ATM networks use the standard layered routing model. This means that PNNI is used for ATM routing (of SVC requests) among the switches and a routing protocol like OSPF is used on the routers to compute paths for layer-3 packets. However a routing protocol like OSPF has no knowledge whatsoever of the internals or topology of the ATM network. As far as OSPF is concerned an ATM subnet may appear as a point-to-point link (RFC 1483 PVC), an emulated LAN (LANE) or an LIS (RFC 1577). Routers that need to talk to each other across an ATM network may use any one of these techniques to establish connectivity. But these techniques may become unmanageable as the ATM network becomes larger and more routers are placed at the edge.

As the name implies, PNNI augmented routing (PAR) augments the IP routing protocol (for example OSPF) on the ATM-attached routers with an instance of PNNI. When PAR is used, all routers run one or more IP routing protocols (business as usual), and use the results for forwarding of IP packets. All ATM switches run standard PNNI Phase I. Those routers with ATM interfaces will also participate in the operation of the ATM PNNI routing protocol. The information obtained from the PNNI allows these routers to locate other routers in the ATM network, and provides these routers with detailed internal knowledge of the state of the ATM network. This eliminates the need for manual configuration of PVCs or the requirements to use LANE, RFC 1577 or NHRP to establish SVCs with other routers on the ATM network. This simplifies the initialization and network management for these routers and allows more efficient maintenance of SVCs between these routers across the ATM network. Figure 57 on page 149, illustrates a network of routers all running OSPF. The ATM-attached routers are also running PNNI.

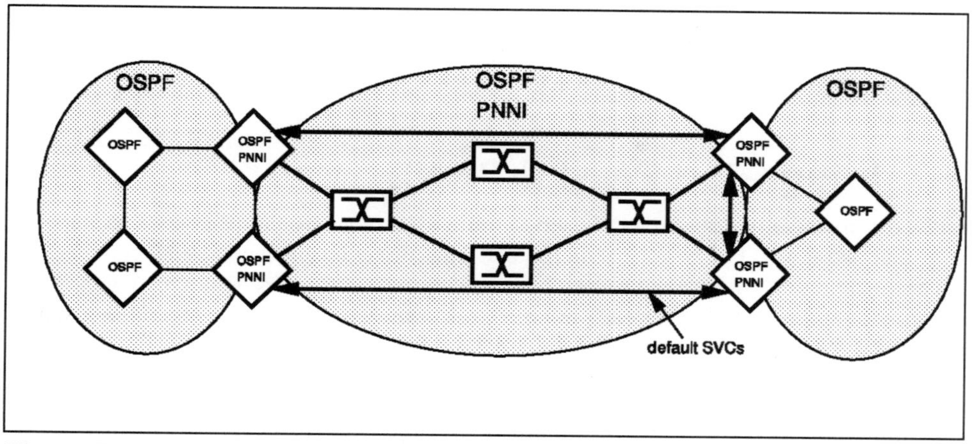

Figure 57. *PNNI Augmented Routing*

The extensibility features of PNNI are used to advertise IP-specific information using type/length/value (TLV) encoding. This allows IP-specific information to be transmitted by routers, for the benefit of other routers, but be ignored by the ATM switches that are running PNNI Phase 1. Routers are advertised in PNNI as *restricted transit* nodes. This allows the router to originate or terminate an SVC without being called upon to be an intermediate node of an SVC. These features allow the routers implementing PAR to be fully compatible with ATM switches that implement only PNNI Phase 1. The ATM switches are not required to have any knowledge whatsoever of IP routing.

Routers running PAR participate as regular PNNI nodes, exchange PNNI hellos with neighboring ATM switches, and exchange regular PNNI PTSEs throughout the peer group. PAR routers announce normal PNNI information such as links to neighboring ATM switches, metrics on these links, and the node ID and ATM address of the router. In addition, the PAR routers use the extensibility features of PNNI to announce, in a way that normal ATM switches will ignore, router-specific information such as the internetwork level protocol supported (for example, IP), router ID, the IP routing protocol in use (for example, OSPF or RIP), and routing-protocol specific information, such as the OSPF area.

PAR may be thought of as an optimization of layered routing. It allows routers, including route servers, to find each other across the ATM network, and thereby facilitates auto-configuration for SVCs between routers. It also ensures that a reasonable set of default SVCs is established automatically as part of network initialization. This reduces the need to buffer or discard IP packets while waiting for NHRP queries to be answered. It provides automatic adjustment of the SVCs between routers as routers join or leave a PNNI peer group, and it provides automatic adjustment of the SVCs based on transient features such as traffic load.

PAR allows existing IP routing protocols to operate essentially unchanged over a combination of legacy and ATM networks. PAR can be extended to support other internetwork layer protocol suites such as IPX, DECnet, and Appletalk.

6.7 Integrated PNNI (I-PNNI)

I-PNNI is a single routing protocol used to route SVC requests and IP layer-3 packets over a network of ATM switches and routers. This single instance of I-PNNI runs on ATM switches, ATM-attached routers and legacy-attached routers. Routers no longer run a separate routing protocol for each internetwork layer they are supporting but rather one routing protocol: I-PNNI. The motivation behind I-PNNI is based on the belief that IP and other routed protocols can benefit from the unique properties supported by PNNI such as QOS-based path selection, auto-configuration, scalability and topological flexibility. And because routers are running I-PNNI they possess a view of the real ATM network topology. In addition it is quite clear that existing routing protocols such as OSPF and RIP could add complexity if not hamper the operation of IP over large switch-dense networks.

6.7.1 I-PNNI Operation

Switches and routers running I-PNNI behave just like normal PNNI nodes. Routers and switches are configured with node identifiers and peer group identifiers. (This includes routers attached to legacy LANs and other networks.) Standard PNNI hellos are exchanged between switches, switches and routers and, router and routers. Peer groups can be formed consisting ATM switches, routers and switches or just routers.

Figure 58 on page 151, shows a topology of switches and routers all running I-PNNI. Notice in peer group C the presence of a route server and edge devices that make up a virtual or distributed router. The route server can run I-PNNI as can conventional routers and switches. The edge devices do not run I-PNNI but instead a standard ATM ELAN protocol such as LANE or MPOA.

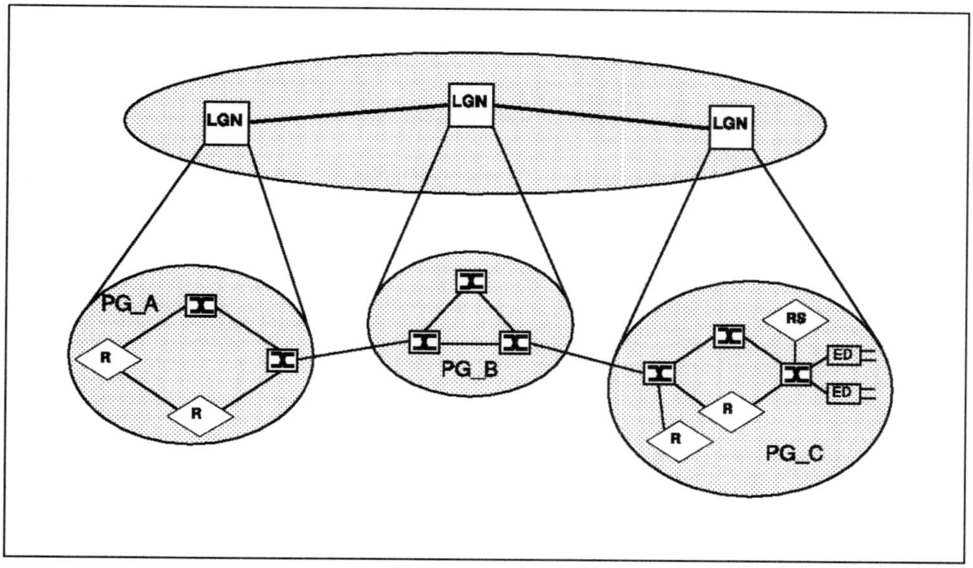

Figure 58. I-PNNI Topology

Router and switches will announce local topology in PTSPs and exchange this information with their neighbors. IP addressing information will be encoded in a TLV-encoded PTSE that is separate from the normal PTSE that contains ATM address reachability. Thus the advertisements for ATM and IP address reachability will be kept separate. ATM switches will announce reachability to the ATM addresses that they can reach, based on prefixes of 20-byte ATM private addresses. Similarly, routers announce reachability to those IP addresses that they can reach, based on prefixes on 4-byte IP addresses.

ATM switches do not know anything about IP routing or addressing. I-PNNI utilizes techniques defined in PNNI Phase I which explain how nodes are to handle information that they do not know about. In this case a TLV containing IP addressing information will be left alone by the ATM switch. The ATM switch will store the information as a PTSE in its topology database and forward it in a PTSE to adjacent nodes in the peer group but it will not use the information in performing a route computation for an SVC request.

Normal routers cannot be used as a transit system for SVCs. However it is certainly possible for SVCs to originate and terminate in a router. To make this limitation known to other members of the peer group (in particular the ATM switches) the routers will announce themselves as restricted transit nodes. This is a topology attribute which indicates that this node cannot forward SVC requests. Thus when a router or switch is generating a DTL, it will never include a restricted transit node as an intermediate node in the source routed path.

6.7.2 I-PNNI IP Address Reachability

Reachability to IP subnets/hosts can be advertised in several ways with I-PNNI. First of all it is necessary to allow routers to announce that they can directly reach particular IP addresses (corresponding to host routes, subnets, or summary routes). This is done in the same manner as other IP routing protocols (specifically OSPF and I.IS-IS), by announcing a prefix of an IP address. This information is included in PTSEs initiated by a router. When a router announces direct reachability to a particular address prefix, this implies that the router is directly attached to the associated IP hosts or subnet. In order to deliver the packet to the associated destination(s), the IP packets may be transmitted to the router directly.

The second type of IP reachability that can be advertised by I-PNNI routers is query service. This reflects the case where the I-PNNI router who is advertising reachability to an IP address prefix may not provide the the most optimal or direct path to the specified IP address prefix. More accurate information can be obtained by sending an NHRP query to the advertising router. If a router announces *reachability* to a particular IP address, this implies that the router is capable of reaching the associated IP hosts or subnets, but is not necessarily on the optimal path to the associated host or subnet. Thus a more optimal route may be achieved if an NHRP query is first used to determine the optimal ATM address to use to reach any particular IP address which matches the advertisement.

Query reachability is applicable in the case where a route server is supporting edge devices or ATM-attached hosts. In this case, optimal routing to a specified IP address may require that an NHRP address resolution query should be directed to the route server (which is announcing reachability to the destination subnet), which will in turn will reply by providing the ATM address to use for the best path to that host or edge device. This is also implied by the fact that route servers running I-PNNI (or any other routing protocol) should not advertise specific host routes. Thus to obtain the ATM address of a specific destination IP address a NHRP query is required.

An example of this would be the route server in peer group C in Figure 58 on page 151. It will announce within its own peer group that the subnet(s) that it supports are query reachable. Therefore any other router who wishes to forward a packet to the IP subnet advertised by the route server will first send a NHRP query. The route server will return a NHRP reply that contains the ″nearest″ ATM address to the destination.

Query reachability is also useful in hierarchical networks. Here a single real physical system (the peer group leader) will be transmitting PTSEs describing the capability of a logical group node (LGN). Thus, the PTSEs associated with the LGN may advertise summary reachability to multiple IP systems using one or more IP address prefixes.

However, the optimal ATM address to use to reach any possible IP address might not be advertised in the summary reachability advertisement, and instead the optimal ATM address to reach any one particular IP address may be determined by using an NHRP query.

To support query reachability advertisements all I-PNNI knowledgeable systems must be capable of supporting the NHRP server function.

The third type of reachability supported is *In Care Of* addresses. There may be some cases where a router or route server wants to advertise reachability on behalf of another system. For example, a router may know of a major server which is either attached to a virtual network served by the router. Similarly, a logical group node may want to advertise the address of a major server which is included in a lower level peer group. In this way, for those "popular" addresses which are frequently contacted by other systems, it is possible to avoid the need for a query/response. In order to announce "in care of" reachability to a particular IP address or address range, the router advertises the ATM address that should be used to deliver IP packets to the specified destination or group of destinations. In this case the router advertises an IP address prefix plus a single ATM address. In order to deliver packets to any IP destination which matches the specified address prefix, an SVC may be set up to the specified ATM address.

6.7.3 Route Computation

I-PNNI nodes will compute DTLs for ATM SVCs. This is true for switches and true for routers originating an SVC. Routers are restricted transit nodes and thus are precluded from forwarding SVC requests. For best effort IP traffic, I-PNNI uses hop-by-hop routing. This implies that the routers within a peer group must perform a consistent route computation, in order to protect against routing loops. (This requirement is the same as occurs with any link state routing protocol, including OSPF.) For this reason I-PNNI specifies the standard route computation to be used for best effort hop-by-hop IP packets. A standard route computation for best effort IP packets consist of a simple Dijkstra computation on the administrative weight.

For SVC requests, I-PNNI can reroute around failed components using the crankback and alternate routing mechanism. It does not enable an ATM SVC to dynamically switch to an alternate route once the call is established.

For IP packets, I-PNNI can reroute around failed components just like any other IP routing protocol. This is because IP is connectionless and the route computation is performed on a hop-by-hop basis.

6.7.4 I-PNNI and Broadcast LANs

Routers running I-PNNI will almost certainly attach to broadcast LANs such as Ethernet. However ATM is based on point-to-point media (VCCs) and does not support broadcast LANs. In this case the developers of I-PNNI borrowed a technique from OSPF in which the concept of a designated router and pseudonode is used to represent the broadcast LAN in the topology database. This capability will only run in I-PNNI routers attached to broadcast LANs.

6.7.5 I-PNNI Compared to MPOA

MPOA defines a set of behaviors and services that enable a host to interact with routers or route servers over an ATM subnet, and for distributing routes from a route server to an edge device forwarder. MPOA is not a routing protocol and does not specify any specific routing protocol to be used. MPOA is compatible with a wide choice of routing protocols including I-PNNI.

I-PNNI is a routing protocol for ATM, for IP, and potentially for other internet level protocols. However, I-PNNI does not define host behavior, and is compatible with a range of host behaviors. Similarly, I-PNNI does not require, but is compatible with the separation of route server and edge device forwarding functions between different physical devices.

Figure 59, attempts to illustrate the scope of MPOA and I-PNNI. MPOA services are confined to hosts, edge devices and route servers. I-PNNI is a routing protocol that runs on routers, route servers and switches. The two do not overlap and in fact are entirely compatible.

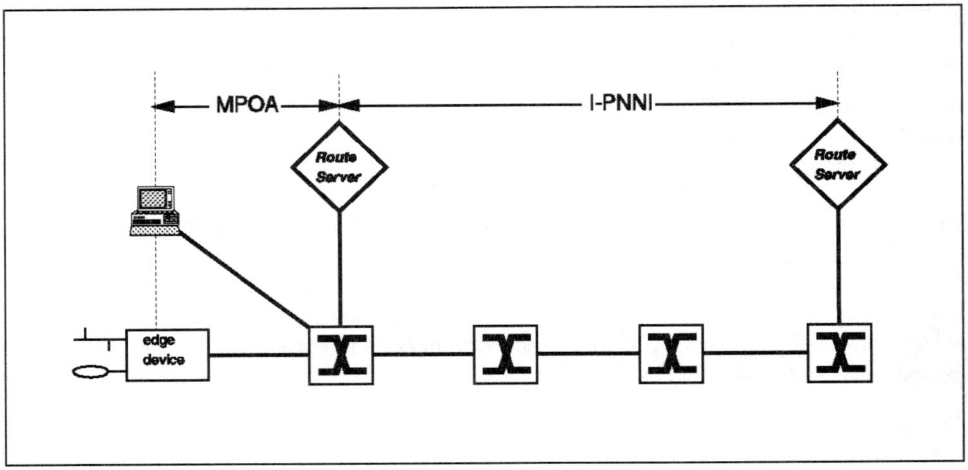

Figure 59. *MPOA and I-PNNI Scope*

6.7.6 I-PNNI Summary of Features

Integrated PNNI is still an early work in progress but its salient characteristics can be summarized by the following:

- Single protocol for IP and ATM routing.
- Can be extended to support other layer-3 internetworking protocols.
- Can run on routers attached to ATM and legacy networks.
- Multiple levels of hierarchy.
- Requires NHRP Server on I-PNNI knowledgeable nodes.
- Several types of reachability are defined.
- Fully compatible with PNNI Phase I.
- Compliments MPOA where I-PNNI would run on MPOA Route Server.
- I-PNNI is a QOS-sensitive routing protocol. Enhancements to the current RSVP/IntServ model or some other flow identification technique must be introduced so that IP can make use of QOS-sensitive routing.
- Works over broadcast LANs.
- Targeted for large switch-dense networks.
- PAR and I-PNNI are work items in ATM Forum PNNI SWG.

Chapter 7. Switched Virtual Networking Architecture

Switched Virtual Networking (SVN) is a comprehensive approach for building and managing switch-based networks. SVN is IBM's ATM strategy and combines the virtues of LAN switching, bridging, routing, ATM switching, and other switched services.

7.1 SVN Components

The ultimate goal of SVN's switch-centric approach is to utilize ATM for the core backbone and then exploit this by moving existing infrastructures and function, such as routing, SNA, TDM, bridging and voice switching to the periphery. In addition, the routing function is being moved out of the data path as shown in Figure 60 on page 158. When needed, a server function is queried for the ATM address of a partner, then a direct *cut-through* route is established to the partner taking the routing function out of the data path.

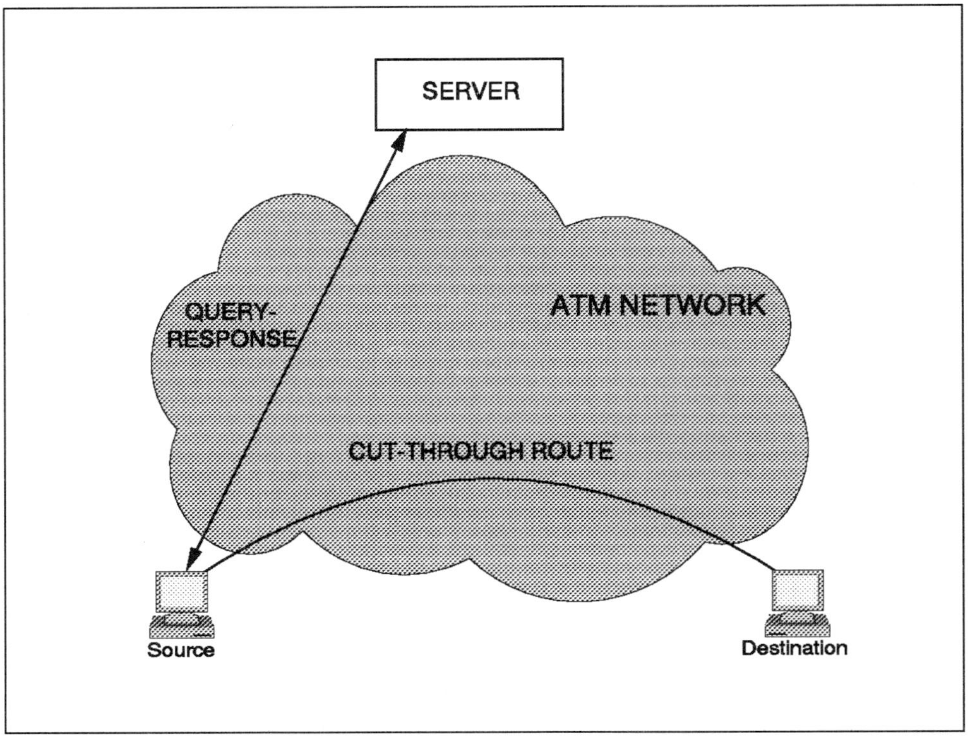

Figure 60. Cut-Through Routes

The key functional elements of the SVN strategy for both campus and wide area environments include:

- Periphery switching
- Backbone switching
- Network management

Figure 61 on page 159 shows the major components of SVN.

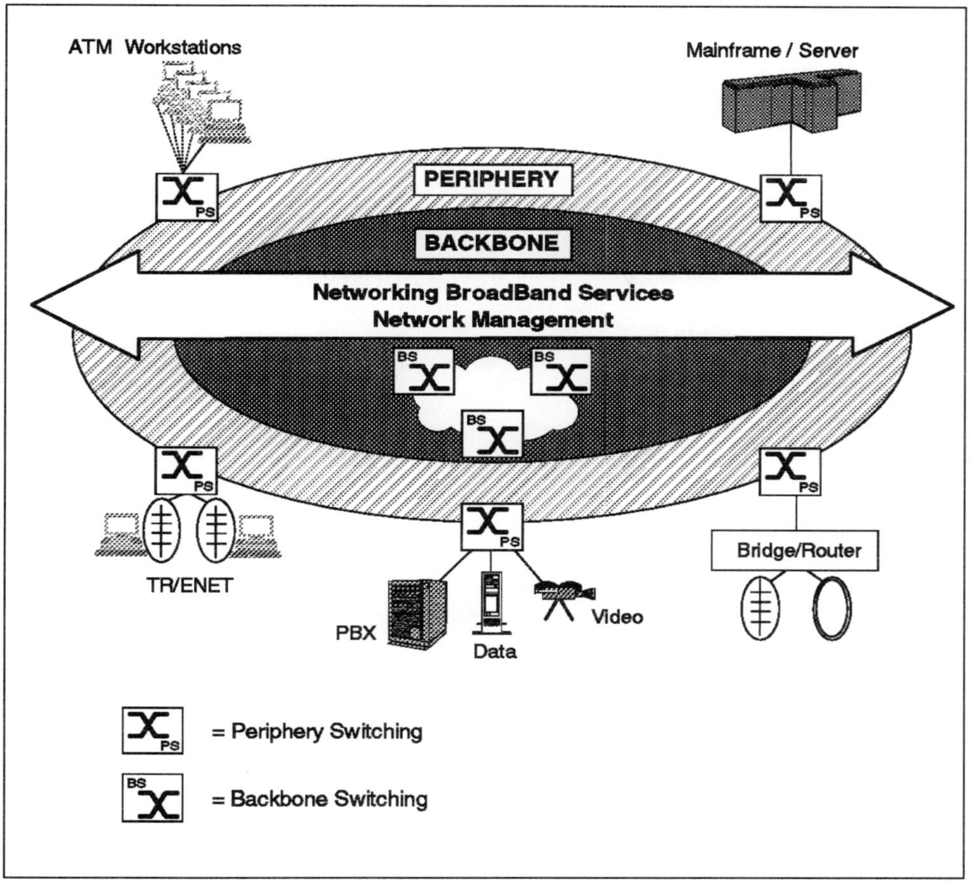

Figure 61. Switched Virtual Network (SVN) Components

7.1.1 IBM Networking BroadBand Services

SVN is based on a single comprehensive architecture, NBBS (see 19 on page 231); specifically it is designed to address the unique needs of very high-speed switched networks. NBBS has been extended beyond its original focus on the wide area specific functions of Access, Transport, and Advanced Network Control Services to embrace the LAN environment with MultiProtocol Switched Services (MSS).

7.2, "IBM Networking BroadBand Services" on page 162 explains the details of NBBS, its operation, and how it relates to the standards, both developed and emerging.

7.1.1.1 Backbone Switching

In a network backbone, ATM switching provides high-speed connectivity, reliability, and quality of service between periphery switches, thus enabling the consolidation of multiple traffic types. Without efficient congestion and flow control, high availability, sophisticated network control, dynamic user group management, and effective traffic management, today's bandwidth demands and tomorrow's multimedia applications cannot be well-served or perhaps even satisfied at all. At the same time, any products that enable an ATM backbone must also conform to the standards being developed by standard bodies (for example, the IETF) and the ATM Forum (UNI, P-NNI, B-ICI). It is clear that many of these sophisticated features are not required and would not be cost-effective to provide in the periphery of a network.

7.1.1.2 Peripheral Switching

Periphery switching enables the access of both new and existing equipment to the high-speed switched backbone in either the campus or wide area environment. By extending the switching function all the way to the edge of the network, for example, implementing the function in a LAN switch, any end station can appear to be a single hop away from any other end station. In other words, one is able to build a direct virtual circuit (VC), as shown in Figure 15 on page 27, rather than using the hop-by-hop approach typical of a multiprotocol router implementation shown in Figure 13 on page 26. This approach is called *one-hop routing*.

As described in 1.2.2, "The Changing Role of Routing" on page 5, the aim is to create a protocol-independent network core. Hence, to take this approach to its limit, we can move the switching function all the way out to the client workstations. This enables *no-hop routing* to take place.

IBM Multiprotocol Switched Services: IBM MSS provides peripheral switching services. MSS consists of distributed routing, enhanced LAN emulation, broadcast management, and VLAN support. Together, they can be viewed as a new access service of NBBS that provides NBBS functions all the way to the end station. Of key importance is the fact that MSS is based on today's LAN, LAN emulation, and multiprotocol standards.

7.3, "IBM Multiprotocol Switched Services" on page 171 gives an overview of Multiprotocol Switched Services (MSS), and how it relates to the standards, both developed and emerging.

7.1.1.3 Network Management

The focus for SVN leadership in network management is centered around the following key strategic points:

- Manage all elements of the switched virtual network
- Deliver added value applications and functions that address both physical and logical views of the network
- Address virtual LAN management
- Provide attractively priced application packages with integrated installation procedures
- Provide management support of both IBM and non-IBM hardware
- Implement management support on multiple platforms, both IBM and non-IBM

A common customer requirement is to be able to view the enterprise network as a set of interconnected physical devices. As network device functions have increased, and as the kinds of network protocols in the enterprise have increased, the relationship between physical and logical resources has become increasingly complex. IBM's strategy is to provide the Network Management System (NMS) operator the ability to view the physical network and tailor the view to meet organizational and ease of use goals. The relationships between the physical and logical views are maintained by the NMS, and the operator can navigate between them.

Virtual LANs provide the means to meet performance and security goals of workgroups, which are independent of how the workstations and servers in a group are attached to the network. The separation of the physical and logical network structure also provides the opportunity to administer moves, adds, and changes from the NMS.

IBM's strategy is to provide management for the different kinds of VLANs: logical LANs created with intelligent hubs, virtual domains created with LAN switches, ATM LAN Emulation, and Multiprotocol Switched Services. VLAN administration is done from a common graphical interface for all kinds of VLANs. Moves, adds, and changes can be accomplished with simple drag-and-drop operations; reconfiguration of the network to implement the change can, in many cases, be handled automatically. Other VLAN management functions include auto-discovery, status monitoring, performance fault, and security.

7.2 IBM Networking BroadBand Services

This section provides a short overview of IBM Networking Broadband Services (NBBS).

7.2.1 Architecture Overview

IBM Networking BroadBand Services (NBBS) is an architecture for providing sophisticated telecommunication services using networks of ATM switching equipment, such as the IBM 2220 Nways BroadBand Switch family.

ATM switching equipment provides a *fast packet-switching* capability that has been adopted by the telecommunications industry as a standard way of supporting new high-speed multimedia telecommunications services. These services are known as broadband integrated services digital network (B-ISDN) services.

NBBS provides a very complete set of functions that can be used to provide both traditional (pre B-ISDN) telecommunication services and the new B-ISDN services, using a common network of ATM switching equipment. The ability of NBBS to support traditional telecommunication services enables ATM switching technology to be deployed today, and the benefits of fast packet transfer to be enjoyed immediately. This is important because there are very few communication devices in wide use that currently incorporate ATM interfaces.

The NBBS architecture provides the facilities described in the relevant ATM (B-ISDN) standards and also incorporates many additional *value-added* functions that enable networks of ATM switching equipment to be deployed more effectively. Some of these value-added functions will ultimately be standardized and when this occurs, the NBBS architecture will evolve to incorporate the new relevant standards.

7.2.1.1 Basic NBBS Terminology

NBBS provides services that enable networks of ATM switches to be used effectively. In order to describe these services and how they operate, it is first necessary to introduce some terminology.

Consider the network illustrated in Figure 62 on page 163. This network is comprised of multiple ATM switches that are called *subnodes*. Subnodes are connected to each other via communication *links*. Groups of subnodes that are packaged together and managed as a single entity are called a *node*. Within each node, there is a *control point* that is responsible for controlling the operation of all the subnodes that form the node. Services that are provided using NBBS are made accessible to external communication devices via *access link* interfaces.

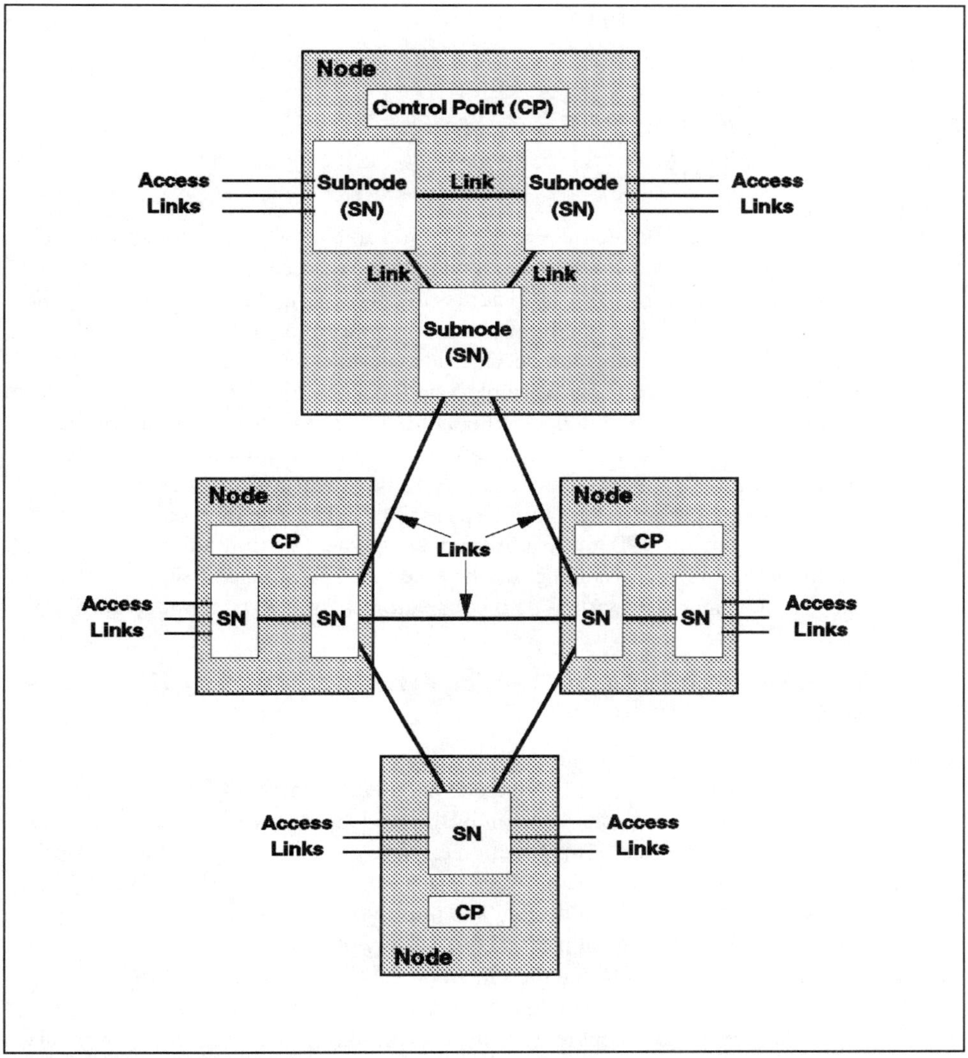

Figure 62. A Network of ATM Switches Using NBBS

Notes:

1. The network diagram represents the control point as a separate entity (even without any connection to subnodes in the node). Actual implementation of the control point may be distributed between the subnodes that form a node.
2. Current NBBS implementations (such as the IBM 2220 Nways BroadBand Switch) have only a single subnode in each node.

3. Links between nodes are also known as *trunks*.

4. Access link interfaces are also known as *ports*.

7.2.1.2 Services of an NBBS Network

The services that NBBS provides can be divided into three major areas:

Transport Services

Transport services enable the nodes in a wide area network to support diverse traffic types over links operating at a wide range of speeds. It provides for transmission scheduling of traffic across the network and does queueing by the relative delay properties of circuit emulation, real-time, non-real-time, and nonreserved traffic. Transport services also permit links to be configured to transmit both ATM calls and variable length packets. This is an attractive option for situations in which only data is being transmitted across all or part of the network (for example, for frame relay). In addition, it supports a preemptive service (a higher-priority packet can preempt a lower-priority packet) for variable length packets over slower speed (T1) links. The implication is that high-priority traffic will not be impeded, even when the links are not fast enough to support all the traffic at once. In the case of ATM, transport services manage the virtual channel and virtual path connections that have been set up across the network.

The transport services are not used directly; instead, they are used by the access services.

Access Services

Access services provide a like image to an attaching network; for example, a frame relay TE would connect to an NBBS node as if it were simply connecting to an FR bearer service. This function ensures the support of existing networks without modifications to them. Access services provide connection setup for point-to-point, point-to-multipoint, and multipoint-to-multipoint connections. Because of the ability to support a wide variety of attaching devices, the migration to ATM can be handled in an orderly fashion according to the required schedule. They enable a wide range of different types of communication equipment to get access to the common infrastructure provided by the transport services.

Network Control Services

Network control services provide the facilities that are needed to ensure that the transport and access services operate reliably, efficiently, and as automatically (that is, without manual intervention) as possible.

These advanced functions address the new and unique problems of managing a high-speed network comprised of low bit error rate links that must support multiple traffic types requiring guaranteed service levels. Advanced network control services provide directory services, path selection, congestion control, traffic management, topology services, and multicast services in order to allocate, control and manage the network resources.

Some highlights of network control services include the following:

- Congestion Control using the Forum-compliant *dual leaky bucket* technique. Before traffic can enter the network, a traffic contract is established, which specifies the maximum packet/cell rate to which the user must adhere. At the entry point into the network, traffic is policed and if it exceeds the contract, it is tagged as discard eligible. Once a path is chosen for traffic from a given source, bandwidth is allocated for that traffic, then the packets/cells are permitted to traverse the path as long as there is no congestion detected. Once congestion is detected within the network, the tagged packets are discarded in order to maintain the traffic flow that has been guaranteed. In addition, traffic shaping is used to reduce the rate at which traffic enters the network, thus reducing the bandwidth requirements.
- Bandwidth Management can result in significant (50% or more) savings on the cost of wide area links. Algorithms that analyze the traffic according to its burstiness and required quality of service are employed to allocate only the bandwidth that is needed to meet the QOS requirements, as opposed to resorting to peak allocation. This bandwidth determination is made at the time of connection setup. Bandwidth adaptation is then employed over the life of the connection to monitor the bandwidth utilization and adjust it within certain allowable bounds.
- Multicast Service provides for the establishment and maintenance of the membership of end users in multicast groups (for example, a videoconference). It allows a user to establish a connection to a group regardless of whether or not that user is a member of the group. These group connections can be point-to-point, point-to-multipoint, or multipoint-to-multipoint.

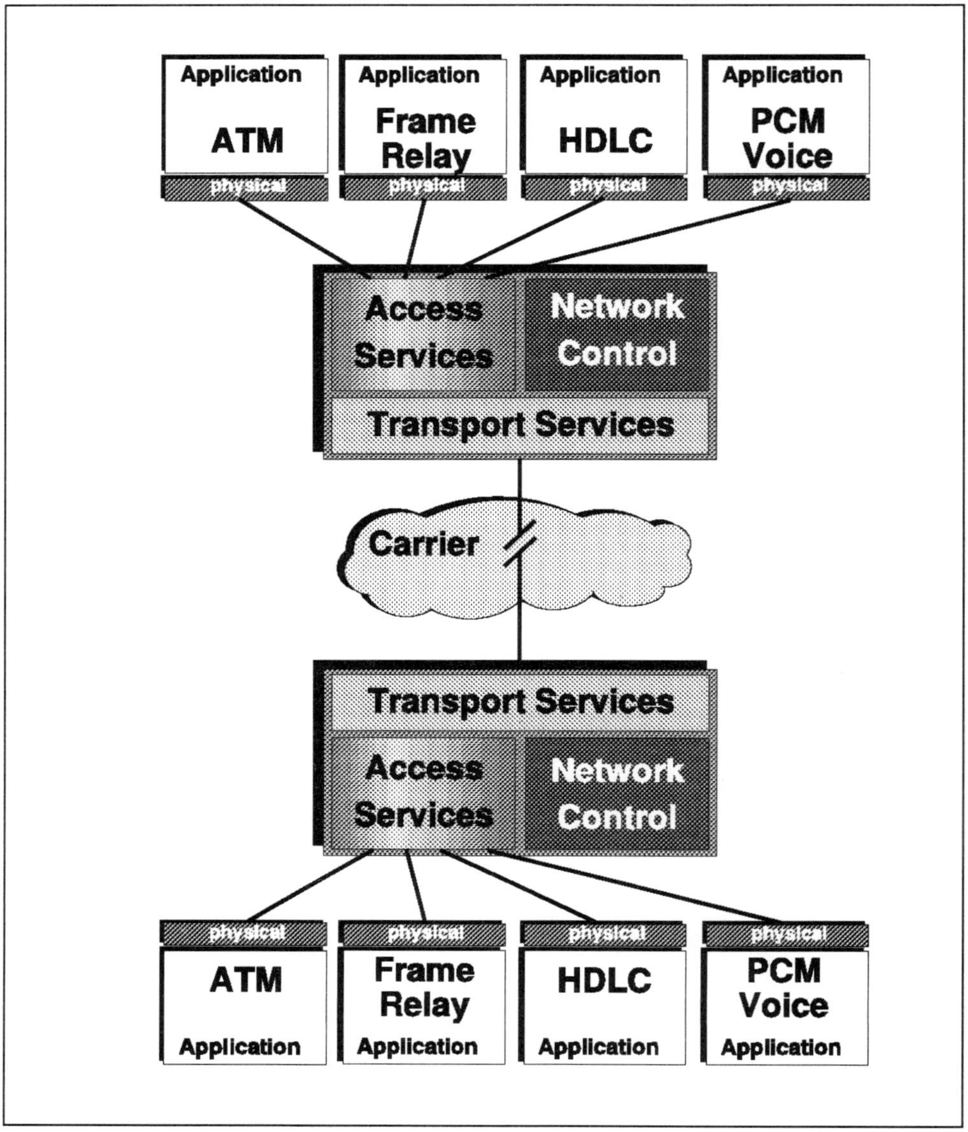

Figure 63. *Services of an NBBS Network*

Together, the transport services, access services, and network control services provide the capability to support communication between many different types of communication devices through a common network infrastructure. This is illustrated in Figure 63.

7.2.1.3 NBBS Architecture Components

We will describe the operation of NBBS by dividing the architecture into various components, each responsible for an element of the architecture. These components and their relationship are illustrated in Figure 64.

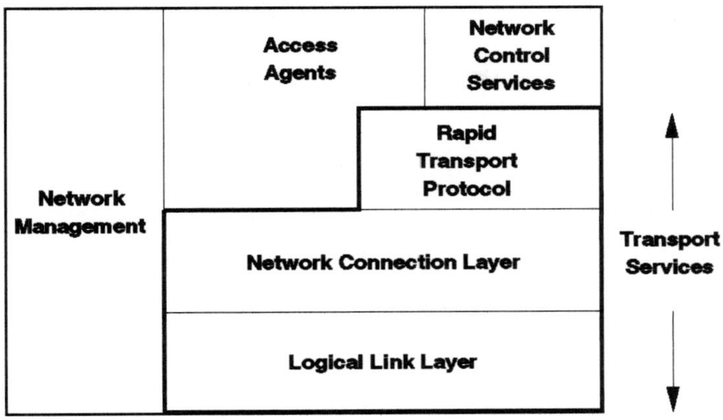

Figure 64. *NBBS Architecture Components*

The logical link layer, network connection layer, and transport protocols together provide NBBS transport services. The access agents are responsible for providing NBBS access services. The network control services and network management provide for NBBS network control.

The roles of each component are:

Logical Link Layer

The logical link layer is responsible for the transport of packets across links between *adjacent* subnodes.

The functions provided by the logical link layer are used by the network connection layer.

Network Connection Layer

The network connection layer is responsible for providing *network connections*. Network connections are virtual pipes that are used to transfer packets between subnodes. Network connections are able to traverse multiple subnodes and nodes. That is, the endpoints of a network connection need not be in adjacent subnodes or nodes.

The network connection layer transfers packets on a best-effort basis and does not guarantee reliable and error-free delivery of packets. If reliable, error-free delivery of packets is required, higher-layer protocols have to be used that implement the necessary functions like error checking, retransmission, segmentation, and reassembly of large messages.

The network connection layer uses the functions provided by the logical link layer. The functions provided by the network connection layer are used by the network control services and access agents either directly or in conjunction with a transport protocol.

Transport Protocols

Transport protocols provide various functions that enable network connections to be used more effectively by the network control services and access agents. Transport protocols are required where additional capabilities are needed beyond those provided by a *raw* network connection. For example, where it is necessary to send a large message over a network connection, a transport protocol can be used to segment the message into smaller packets acceptable to the network connection.

Access Agents

Access agents provide the interface between external devices and an NBBS network via access links. Access agents enable a wide variety of external devices to use the services provided by NBBS. They also provide a framework that will enable new types of devices to be supported in the future without impacting the other parts of the NBBS architecture.

Access agents use network connections provided by the network connection layer in conjunction with transport protocols where appropriate. Access agents also interact extensively with the network control services.

Network Control Services

The network control services are responsible for various facilities that are used to control, allocate, and manage the resources of a network on a real-time basis.

Network control services in NBBS are distributed throughout the network and rely on messages that are transported across network connections provided by the network connection layer using transport protocols. The network control services also interact extensively with access agents.

Network Management

Network management provides network operators with the various facilities that are needed to configure, operate, and maintain a network on a day-to-day basis. This includes facilities for monitoring the performance of the network, accounting for its usage, and resolving problems.

Network management interacts with all of the other components in the NBBS architecture.

7.2.1.4 Traffic Management

As described earlier, NBBS is able to support a diverse range of high-speed multimedia telecommunication services. This wide range of services is provided using a common network infrastructure based on networks of ATM switching equipment.

However, the resources of any network are finite and therefore there must be an efficient mechanism for sharing these resources among the various services being supported by the network. In particular, the main resource whose capacity must be efficiently managed is bandwidth.

The NBBS architecture uses a concept called *equivalent capacity* as the basis of a sophisticated *bandwidth management* system.

Equivalent Capacity

For each service supported by NBBS, an equivalent capacity (equivalent bandwidth) is derived to represent the capacity requirements for that service. The equivalent capacity is based on a statistical characterization of the traffic requirements for a service, specifically:

- The peak bandwidth.
- The mean bandwidth.
- The mean burst duration.

In addition to the traffic requirements for a service, the calculation of equivalent capacity is based on quality of service for that service (particularly the packet loss requirements) and the available buffer capacity within the network. This is to ensure that sufficient capacity is associated with a service so that a guaranteed quality of service can be provided for the service.

For many services, particularly variable-bit-rate (VBR) services, equivalent capacity enables much more efficient allocation of network resources than other schemes that always base capacity on the peak bandwidth requirements for a service. This efficiency does not compromise the ability of NBBS to offer a guaranteed quality of service.

Bandwidth Management

NBBS incorporates a sophisticated bandwidth management system that uses the equivalent capacity concept previously described to efficiently manage the bandwidth resources of a network of ATM switching equipment.

Major features of the NBBS bandwidth management capabilities include:

Bandwidth Reservation

Bandwidth reservation is a function provided by NBBS to set aside or *reserve* bandwidth resources within a network to support a particular service. This ensures that a service is provided only if sufficient capacity is available within the network to support the bandwidth needs of the service.

NBBS bandwidth reservation is based on the equivalent capacity concept described previously. For each service, an equivalent capacity is calculated, and this capacity is used to reserve bandwidth resources on the various links traversed by the service.

Dynamic Bandwidth Allocation

The allocation and reservation of bandwidth capacity is performed dynamically by NBBS.

For new permanent services, NBBS dynamically allocates (reserves) bandwidth when the connection is established.

For services that are switched or semipermanent, NBBS allocates bandwidth to the service only when the service is actually in use. At other times the bandwidth reserved for the service is made available for use by other services.

While a service is active, NBBS allows the bandwidth allocated to that service to be dynamically *adjusted* over time. The bandwidth adjustments can be initiated from several sources, including an automatic *bandwidth adaptation* facility described below.

Bandwidth Adaptation

NBBS does not require the capacity requirements for a service to be accurately known when the service is established. Instead, NBBS provides a sophisticated bandwidth adaptation capability that continuously monitors the capacity required to support a service and dynamically adjusts the bandwidth reservation associated with the service.

7.2.1.5 Quality of Service (QOS)

NBBS is able to provide a *guaranteed quality of service* for a diverse range of high-speed multimedia telecommunication services. The quality-of-service guarantee is provided by careful allocation of network resources and ensures that each new service does not impact the quality of service delivered to other existing services.

To understand the many functions NBBS provides that contribute to a guaranteed quality of service consider the steps that occur when a new service is established using NBBS:

1. The equivalent capacity requirements are determined for the service.
2. A path is selected through the network that can meet the capacity and quality-of-service requirements for the service.
3. Network resources are reserved along the path that the service will take.

In the event that a suitable path is not available to meet the capacity and quality-of-service requirements for a service, a *path preemption* capability can be used to force an existing lower-priority service to change paths so that the new service can use the path that meets its capacity and quality-of-service requirements.

Once the service is operational, additional functions are used to ensure that the service does not impact other services and that the quality of service required is achieved.

Congestion control functions are used to limit the network resources that a service can use beyond the allocated resources. This is done so that the quality of service of other services is not impacted by congestion that could occur if a service were allowed to consume unlimited network resources.

A prioritization scheme is used to control the use of buffers and the scheduling of transmission of packets within the network. This ensures that services with stringent quality-of-service requirements can meet those requirements.

To meet availability quality-of-service requirements, NBBS provides a nondisruptive path switch function that can be used to maintain services that have high availability requirements despite events that may have otherwise caused the service to fail.

7.3 IBM Multiprotocol Switched Services

MSS uses the following standards-compliant technologies:

- ATM Forum Compliant LAN Emulation
- Classical IP and ARP over ATM (RFC 1577)
- Layer 3 Routing
- Layer 2 Bridging

MSS also offers the following enhancements to the previously mentioned standard technologies:

- Distributed LES
- Broadcast Manager
- Security

- ELAN Assignment Policies

7.3.1 Enhanced LAN Emulation Component

IBM MSS Release 1 contains the following extensions to ATM Forum LAN emulation:

Distributed LES

A distributed LAN emulation service does not alter the standardized interfaces and operating characteristics of any ATM emulated LAN. The LAN emulation service functions are distributed across multiple platforms for two primary purposes:

- Robustness
- Scalability

A draft contribution (see 17 on page 230) now exists detailing a LAN emulation network-to-network interface (LNNI). This will be implemented once it is agreed upon by the ATM Forum.

Since the LNNI specification is not yet available, backup LE servers function will provide the robustness that LNNI will provide. A LES/BUS instance may be configured to participate in a *redundancy protocol* as either a primary or backup server for a given ELAN. On detecting a failure of the LES/BUS, the LECS will connect to the backup LES/BUS. LECS robustness is achieved by including multiple LECS addresses in the ILMI database. An LEC will connect to a backup LECS if the primary is unavailable.

Figure 65 on page 173 shows the distributed LES function.

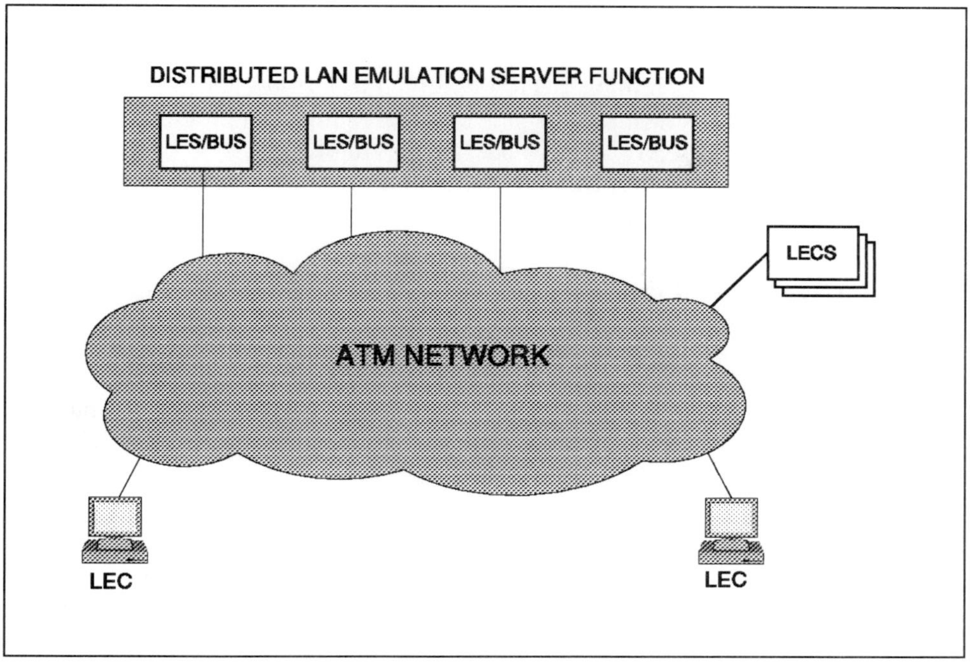

Figure 65. Distributed LES

BroadCast Manager

In an ATM emulated LAN (ELAN), the broadcast mechanism is provided by an entity called the broadcast and unknown server (BUS). Every broadcast frame is sent to the BUS, which then forwards the frame to all LAN emulation clients (LECs) on the ELAN. Clients that function as bridges then forward the frame onto other LAN segments. All end stations receive and process broadcast frames.

The broadcast manager is a value-add extension of the BUS. The principal goals of the Broadcast Manager (BCM) are as follows:

1. Improve overall performance and efficiency by reducing both network traffic and end-station processing overhead associated with filtering spurious frames.
2. Enable practical deployment of larger ELANs. Broadcast management is also useful when ATM connections traverse a WAN, where bandwidth is more price sensitive.

Broadcast Manager will operate independently within each ELAN by intercepting broadcast and multicast frames sent to the BUS. Frames intended for another ELAN must be forwarded using another mechanism, such as an

ATM router or bridge. BCM performs a minimum amount of layer 2 and layer 3 decodes in order to perform broadcast management.

Security

To control ELAN membership, a LES may be configured to validate LE_JOIN_REQUESTs. In this mode, the LES forms an LE_CONFIGURE_REQUEST on behalf of the LEC using information from the LE_JOIN_REQUEST. The LES then transmits this request to the LECS. The LEC is not allowed to join the ELAN without LECS approval.

LECS ELAN Assignment Policies

The LECS supports a flexible LEC-to-ELAN assignment strategy that is aligned to the recent LE Service MIB proposals to the ATM Forum. The LECS can be configured with any number of (*policy, priority*) pairs, where *policy* is any of the policies supported by the LECS, and *priority* gives the priority at which that policy is to be considered. The current policies supported by the LECS are:

- ATM address
- MAC address
- Route descriptor
- LAN type
- Maximum frame size
- ELAN name

The policies correspond with fields that may be present in the LE_CONFIGURE_REQUEST. The only field required is the ATM address.

The LECS is configured with a number of (*policy value, ELAN*) pairs. These pairs represent specific mappings of *policy value* (for example a specific MAC address or route descriptor) to an *ELAN*. Upon recepit of an LE_CONFIGURE_REQUEST, the LECS attempts to assign the requesting LEC to an ELAN based on its prioritized list of policies and the values specified in the request.

Protocol VLANs (PVLAN) could be distinguished by giving each PVLAN a unique ELAN name. The LECS would then be configured to assign by ELAN name at a high priority.

Chapter 8. The Future of IP

IP, which started in the 1960s as a research project, has turned into the most widely used networking protocol today. If IP is to keep its popularity, it will need to keep evolving as applications and networks evolve. The major changes to IP being planned are described in the following sections, including plans that enable IP to make use of ATM networks.

8.1 IP Next Generation Protocol

The following section describes the IP next generation protocol. The recommendation for this was issued in January 1995 as RFC1752. RFC1752 describes the requirements for the next generation of IP, which is now known as IPv6 (Internet Protocol Version 6), and includes RFC1833, which is the complete specification for IPv6, and RFCs 1884, 1886 and 1887, which deal with addressing for IPv6.

8.1.1 Next Generation IP. Why?

An undertaking as large as replacing the underlying protocol used by the Internet is something that should not be taken lightly, but the current Internet protocol (IPv4) is rapidly running out of addresses and cannot easily be enhanced to support future needs. The Internet, which earlier carried data for text-based applications, is increasingly being used today to transport multimedia data and even near real-time applications. IPv4 cannot support these applications efficiently. In addition, the user base that IPv4 currently serves has been growing at an exponential rate. In 1987 it was estimated that there would be a need to address as many as 100,000 networks some time in the future. This number will probably be reached in 1996. New estimates speak of the need to address millions of networks in the near future.

Ipv6 solves the Internet scaling problem. It provides a flexible transition mechanism for the current Internet, and was designed to meet the needs of new markets such as nomadic personal computing devices, networked entertainment, and device control. It does this in an evolutionary way that reduces the risk of architectural problems.

Ease of transition is a key point in the design of IPv6. It is not something that was added in at the end. IPv6 is designed to interoperate with IPv4. Specific mechanisms (embedded IPv4 addresses, pseudo-checksum rules, etc.) were built into IPv6 to support transition and compatibility with IPv4. It was designed to permit a gradual and piecemeal deployment with a minimum of dependencies.

IPv6 supports large hierarchical addresses that will allow the Internet to continue to grow and provide new routing capabilities not built into IPv4. It has anycast addresses that can be used for policy route selection and has scoped multicast addresses that provide improved scalability over IPv4 multicast. It also has local use address mechanisms that provide the ability for *plug and play* installation.

The address structure of IPv6 was also designed to support carrying the addresses of other Internet protocol suites. Space was allocated in the addressing plan for IPX and NSAP addresses. This was done to facilitate migration of these Internet protocols to IPv6.

IPv6 provides a platform for new Internet functionality. This includes support for real-time flows, provider selection, host mobility, end-to-end security, auto-configuration, and auto-reconfiguration.

In summary, IPv6 is a new version of IP. It can be installed as a normal software upgrade in Internet devices. It is interoperable with the current IPv4. Its deployment strategy was designed to not have any *flag* days. IPv6 is designed to run well on high performance networks (for example, ATM) and at the same time is still efficient for low bandwidth networks (for example, wireless). In addition, it provides a platform for new Internet functionality that will be required in the near future.

8.1.2 IPv6 Overview

To meet the needs of the Internet and applications that use it, IPv6 was designed to take an evolutionary step from IPv4. Functions that work well in IPv4 were kept; functions that didn't work well were removed. The changes from IPv4 to IPv6 fall primarily into the following categories:

Expanded Routing and Addressing Capabilities

IPv6 increases the IP address size from 32 bits to 128 bits, to support more levels of addressing hierarchy and a much greater number of addressable nodes, and simpler auto-configuration of addresses.

The scalability of multicast routing is improved by adding a *scope* field to multicast addresses.

Anycast Address

A new type of address called an *anycast address* is defined to identify sets of nodes where a packet sent to an anycast address is delivered to one of the nodes. The use of anycast addresses in the IPv6 source route allows nodes to control the path that their traffic flows.

Header Format Simplification

Some IPv4 header fields have been dropped or made optional to reduce the common-case processing cost of packet handling and to keep the bandwidth cost of the IPv6 header as low as possible despite the increased size of the addresses. Even though the IPv6 addresses are four times longer than the IPv4 addresses, the IPv6 header is only twice the size of the IPv4 header.

Improved Support for Options

Changes in the way IP header options are encoded allows for more efficient forwarding, less stringent limits on the length of options, and greater flexibility for introducing new options in the future.

Quality-of-Service Capabilities

A new capability is added to enable the labeling of packets belonging to particular traffic *flows* for which the sender requests special handling, such as nondefault quality of service or real-time service.

Authentication and Privacy Capabilities

IPv6 includes the definition of extensions that provide support for authentication, data integrity, and confidentiality. This is included as a basic element of IPv6 and will be included in all implementations.

The IPv6 protocol consists of two parts, the basic IPv6 header and IPv6 extension headers.

8.1.3 The IPv6 Header

Figure 66 shows the basic structure of an IPv6 packet. In addition to the fixed-length IPv6 header, a number of optional extension headers can also be added. If an optional function is not being used, then its header is not included.

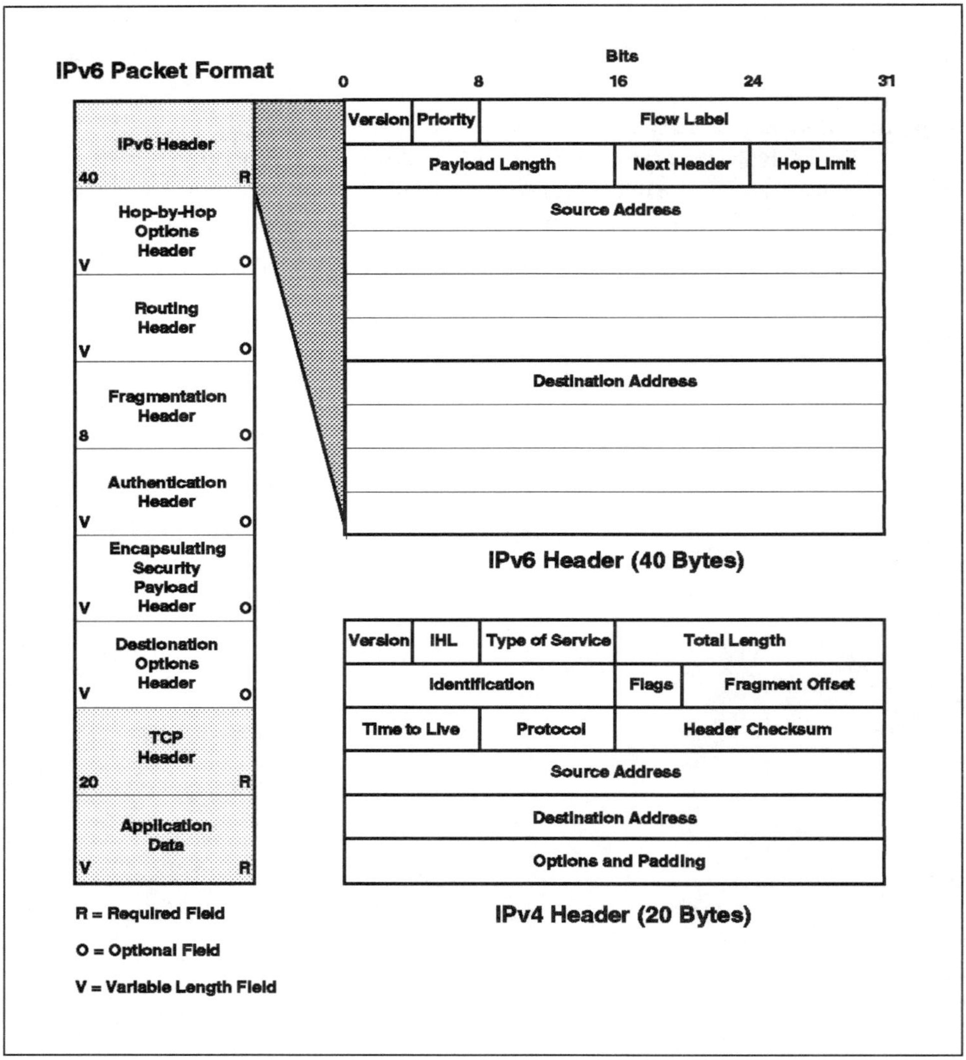

Figure 66. IPv6 Packet Format

8.1.4 IPv6 Addressing

IPv6 addresses are 128 bits long and are identifiers for individual interfaces and sets of interfaces. IPv6 addresses of all types are assigned to interfaces, not nodes. Since each interface belongs to a single node, any of the unicast addresses of that node's interfaces may be used as an identifier for the node. A single interface may be assigned multiple IPv6 addresses of any type.

There are three types of IPv6 addresses. These are unicast, anycast, and multicast. Unicast addresses identify a single interface. Anycast addresses identify a set of interfaces such that a packet sent to an anycast address will be delivered to one member of the set. Multicast addresses identify a group of interfaces, such that a packet sent to a multicast address is delivered to all of the interfaces in the group. There are no broadcast addresses in IPv6, their function being superseded by multicast addresses.

IPv6 supports addresses that are four times the number of bits as IPv4 addresses (128 versus 32). This is 4 billion times 4 billion (2••) times the size of the IPv4 address space (2••).

Unicast Addresses

There are several forms of unicast address assignments in IPv6. These are the global provider-based unicast address, the neutral-interconnect unicast address, the NSAP address, the IPX hierarchical address, the site-local-use address, the link-local-use address, and the IPv4-capable host address. Additional address types can be defined in the future.

Provider-Based Unicast Addresses

Provider-based unicast addresses are used for global communication. They are similar in function to IPv4 addresses under Classless Inter-Domain Routing (CIDR).

Local-Use Addresses

A local-use address is a unicast address that has only local routability scope (within the subnet or within a subscriber network) and may have local or global uniqueness scope. They are intended for use inside of a site for *plug and play* local communication and for bootstrapping up to the use of global addresses.

There are two types of local-use unicast addresses defined. These are link-local and site-local. The link-local-use is for use on a single link and the site-local-use is for use in a single site.

IPv6 Addresses with Embedded IPv4 Addresses

The IPv6 transition mechanisms include a technique for hosts and routers to dynamically tunnel IPv6 packets over IPv4 routing infrastructure. IPv6 nodes that utilize this technique are assigned special IPv6 unicast addresses that carry an IPv4 address in the low-order 32 bits. This type of address is termed an IPv4-compatible IPv6 address.

A second type of IPv6 address that holds an embedded IPv4 address is also defined. This address is used to represent the addresses of IPv4-only nodes (those that do not support IPv6) as IPv6 addresses. This type of address is termed an IPv4-mapped IPv6 address.

Anycast Addresses

An IPv6 anycast address is an address that is assigned to more than one interface (typically belonging to different nodes). A packet sent to an anycast address is routed to the *nearest* interface having that address, according to the routing protocol's measure of distance.

Anycast addresses, when used as part of a route sequence, permits a node to select which of several Internet service providers it wants to carry its traffic. This capability is sometimes called *source selected policies*. This would be implemented by configuring anycast addresses to identify the set of routers belonging to Internet service providers (for example, one anycast address per Internet service provider). These anycast addresses can be used as intermediate addresses in an IPv6 routing header to cause a packet to be delivered via a particular provider or sequence of providers. Other possible uses of anycast addresses are to identify the set of routers attached to a particular subnet or the set of routers providing entry into a particular routing domain.

Anycast addresses are allocated from the unicast address space, using any of the defined unicast address formats. Thus, anycast addresses are syntactically indistinguishable from unicast addresses. When a unicast address is assigned to more than one interface, thus turning it into an anycast address, the nodes to which the address is assigned must be explicitly configured to know that it is an anycast address.

Multicast Addresses

A IPv6 multicast address is an identifier for a group of interfaces. An interface may belong to any number of multicast groups.

8.1.5 IPv6 Routing

Routing in IPv6 is almost identical to IPv4 routing under CIDR except that the addresses are 128-bit IPv6 addresses instead of 32-bit IPv4 addresses. With very straightforward extensions, all of IPv4's routing algorithms (OSPF, RIP, IDRP, ISIS, etc.) can be used to route IPv6.

IPv6 also includes simple routing extensions that support powerful new routing functionality. These capabilities include:

- Provider Selection (based on policy, performance, cost, etc.)
- Host Mobility (route to current location)
- Auto-Readdressing (route to new address)

The new routing functionality is obtained by creating sequences of IPv6 addresses using the IPv6 Routing option. The routing option is used by a IPv6 source to list one or more intermediate nodes (or topological groups) to be *visited* on the way to a packet′s destination. This function is very similar in function to IPv4′s Loose Source and Record Route option.

In order to make address sequences a general function, IPv6 hosts are required in most cases to reverse routes in a packet it receives (if the packet was successfully authenticated using the IPv6 Authentication Header), containing address sequences in order to return the packet to its originator. This approach is taken to make IPv6 host implementations from the start support the handling and reversal of source routes. This is the key for allowing them to work with hosts that implement the new features such as provider selection or extended addresses.

8.1.6 IP Version 6 and ATM

Currently there are several discussions about using IP over ATM. The main problem is the provision of connectionless multicast links over a connection-oriented ATM service. Multicast Address Resolution Server (MARS) provided the support of RFC 1112 style level 2 IP multicast over the ATM Forum′s UNI 3.0/3.1 point-to-multipoint connection service.

8.1.6.1 IP Version 6 versus IP Version 4

Address resolution and address configuration are both a part of the base IPv6 protocol and are located in the network layer rather than in the datalink layer. That is, the Neighbor Discovery protocols that IPv6 uses to perform neighbor and router discovery are an integral part of IPv6, and any mechanism that is used to adapt ATM to IPv6 must deal with the Neighbor Discovery protocols. This is in contrast to IPv4 where the address resolution protocols are not part of the base IP protocols and are part of each individual datalink layer (that is, ARP for broadcast media, ATMARP or NHRP for ATM). In IPv4 new datalink layers could define their own address resolution protocols as necessary (as was done with ATMARP), since this function is left to the datalink. New datalinks could be added without affecting the IPv4 network layer. In IPv6 all datalinks must handle IPv6 Neighbor Discovery packets and use them for address resolution, router discovery and address configuration. Not using Neighbor Discovery would require modifying the IPv6 network layer to accommodate a specific datalink.

IPv6 provides for some extra features not in IPv4. One of these features is address auto-configuration. Any mechanism used to adapt IPv6 to ATM must provide for address auto-configuration since this is expected to be the primary way in which nodes will configure their network layer addresses. IPv6 has also been defined with network layer security features as part of the base protocol. These security features are applied to address configuration and resolution since they are defined at the network layer. Any IPv6 over ATM solution must also take IPv6 security into consideration and preserve the security features built in to address resolution and configuration. Finally, IPv6 includes an address architecture that provides for network layer addresses (specifically link-local and site-local addresses) that are not globally visible. This addressing architecture must also be maintained for IPv6 over ATM.

8.1.6.2 IP Version 6 ATM Implications

The IPv6 protocols currently rely on connectionless broadcast and multicast capabilities of legacy LAN technology to perform functions such as IP-to-physical address resolution. These protocols rely on the physical partitioning of networks to establish the boundaries for the creation of subnets and for routing and address configuration. To adapt the base IPv6 protocols to ATM, the first thing that must be done is to define what an IPv6 subnet and *link* is on an ATM network. That is, how a set of ATM nodes is partitioned into an autonomous group that share common address prefixes, and between which the Neighbor Discovery protocols can be used. Such a *link* should be defined so that all the base IPv6 protocols (including ND) can be run over it with no modifications to end systems or routers, and so that the administration of the network layer software is the same as that on other media.

Once an IPv6 *link* is defined for ATM networks, mechanisms and policies for running IPv6 protocols over the link must be defined. These mechanisms should meet the following goals:

- The concept of the IPv6 subnetwork/link must be maintained.
- The scope of link-local and site-local addressing must be maintained.
- All functionality provided by the current IPv6 Neighbor Discovery protocols must be provided.
- There must be no modifications required to the IPv6 network layer in order to run IPv6 over ATM.
- The Neighbor Discovery protocol semantics must be maintained.
- The resulting protocols and architecture must scale to support arbitrary large networks.
- Nodes must be permitted to join multiple subnets/links.
- Nodes must be permitted to join any subnet/link on the wider subnet without regard to the node's geographic location.
- Nodes on different subnets/links must be permitted (but not required) to establish connectivity directly through the ATM network without going through a router. This is needed so that connections can be established, and IPv6 can make full use of ATM QOS capabilities.
- The link must provide for redundancy and failure of critical, non-ATM resources.

Optionally, any IPv6 over ATM architecture should provide for the highest degree of auto-configuration as the ATM and IPv6 protocols will allow. Ideally, all elements of an IPv6 over ATM network should be self configuring with as little human intervention as possible.

Any protocols developed for IPv6 over ATM should be defined for both current UNI 3.1 and future UNI 4.0 networks. Since it is expected that UNI 4.0 will be widely deployed by the time IPv6 goes into wide use, the new capabilities and features of UNI 4.0 should be taken into account in places where they can improve the functioning of the protocols. However, since initial implementations will be written to existing UNI 3.0/3.1 networks, the protocols must also work under UNI 3.0/3.1.

<draft-ietf-ipatm-ipv6nd-01> (see 15 on page 230) outlines the problems one might face when implementing the things mentioned earlier.

8.2 IP Integrated Services

The idea that ATM may one day become the transport base for the Internet is very attractive, but business reasons indicate that we will not see that day very soon. People are now realizing that the Internet may never move fully to ATM, and therefore, many functions that people hoped would become available with ATM (for example, multicasting or real-time data transmission) must be supported by the Internet itself.

RFC 1633 (see 4 on page 229) describes the following reasons why the Internet architecture must be extended to support real-time IP traffic. The authors write:

> The multicasts of IETF meetings across the Internet have formed a large-scale experiment in sending digitized voice and video through a packet-switched infrastructure. These highly visible experiments have depended upon three enabling technologies:
>
> 1. Many modern workstations now come equipped with built-in multimedia hardware, including audio codecs and video frame-grabbers, and the necessary video gear is now inexpensive.
> 2. IP multicasting, which is not yet generally available in commercial routers, is being provided by the MBONE, a temporary *multicast backbone*.
> 3. Highly sophisticated digital audio and video applications have been developed.
>
> These experiments also showed that an important technical element is still missing: real-time applications often do not work well across the Internet because of variable queueing delays and congestion losses. The Internet, as originally conceived, offers only a very simple quality of service (QOS),

point-to-point best-effort data delivery. Before real-time applications such as remote video, multimedia conferencing, visualization, and virtual reality can be broadly used, the Internet infrastructure must be modified to support real-time QOS, which provides some control over end-to-end packet delays. This extension must be designed from the beginning for multicasting; simply generalizing from the unicast (point-to-point) case does not work.

> Real-time QOS is not the only issue for a next generation of traffic management in the Internet. Network operators are requesting the ability to control the sharing of bandwidth on a particular link among different traffic classes. They want to be able to divide traffic into a few administrative classes and assign to each a minimum percentage of the link bandwidth under conditions of overload, while allowing *unused* bandwidth to be available at other times. These classes may represent different user groups or different protocol families, for example. Such a management facility is commonly called controlled link-sharing. We use the term integrated services (IS) for an Internet service model that includes best-effort service, real-time service, and controlled link-sharing.

The authors of this RFC go on to say:

> We believe that it is now time to begin the engineering that must precede deployment of integrated services in the Internet.

The fundamental service model of the Internet, based on the best-effort delivery service of IP, has been around for about 20 years now. The Integrated Services model is a proposal to extend this model. New components and mechanisms will be added to, and not replace, existing services.

8.2.1 IP Integrated Services Model

The Integrated Services (IS) model proposes two types of service that are targeted towards real-time traffic:

- Guaranteed Service
- Predictive Service

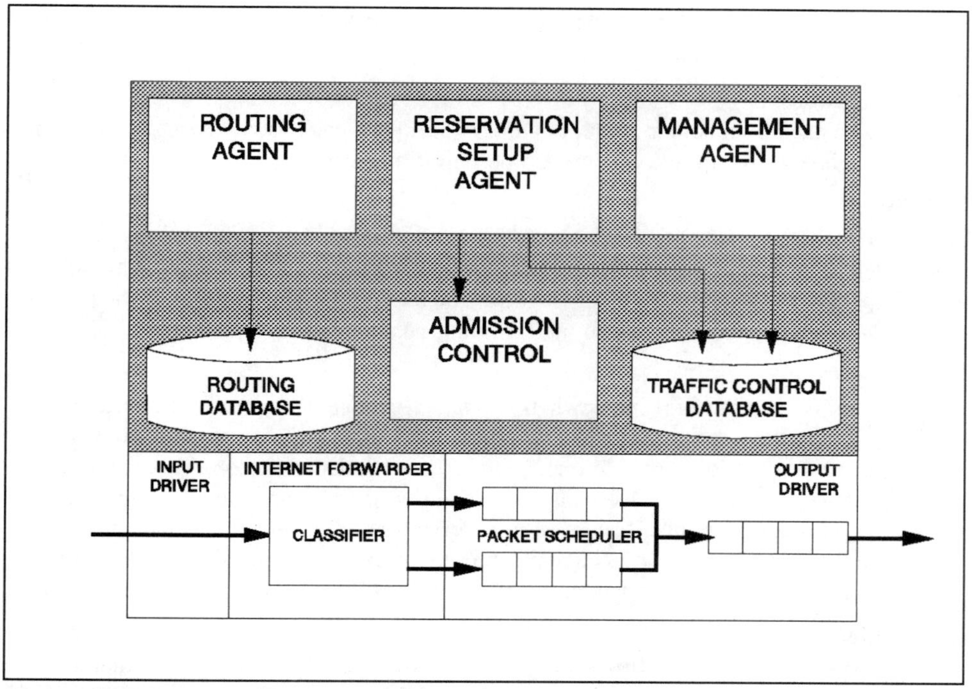

Figure 67. Implementation Model for Routers

These services are integrated with controlled link sharing and are designed to work well with both unicast and multicast traffic.

One of the basic assumptions made is that resources (for example, bandwidth) must be managed, and this implies that some sort of *resource reservation* and *admission control* must be implemented. Although some arguments have been raised against resource guarantees in the Internet, the authors of RFC 1633 (see 4 on page 229) regard resource reservation as an *inescapable requirement.*

Another basic assumption made is that a single IP protocol stack should be used for real-time and non-real-time traffic. This assumption then allows for IS to be added to an existing infrastructure while preserving interoperability. This would allow the single IS model to be deployed everywhere in the Internet, not only an end-to-end deployment supporting end-to-end service guarantees.

8.2.1.1 Reference Implementation Framework

The Reference Implementation Framework for IS consists of four components:

Packet Scheduler
The packet scheduler manages the forwarding of packets belonging to different streams using a set of queues and perhaps timers.

Classifier
For the purpose of traffic control, each incoming packet must be mapped into a class, with all packets of the same class getting the same treatment from the packet scheduler. Choice of class may be based upon the contents of the packet header or some other additional information added to each packet.

Admission Control
Admission control decides whether a new flow can be granted the requested QOS without impacting earlier guarantees. Admission control is invoked at every node along a path to make a local accept/reject decision every time a real-time service request is made. In addition, admission control will enforce administrative policies on resource reservations. This may mean some type of authentication for resources requesting reservations. Admission control will also be reasonable for accounting and administrative reporting.

Reservation Setup Protocol
Some kind of reservation setup protocol will be needed to create and maintain flow-specific states in nodes along the path between endpoints. Resource Reservation Protocol (RSVP) (see 8.3, "Resource Reservation Protocol (RSVP)" on page 189) describes one such protocol.

It should be noted that the previous reservation setup protocol describes a protocol that will set up *flow-specific* states in nodes along the path. This means that the Internet will move away from one of the most important factors in its success. Up until now, flow-specific states were maintained only in the endpoints of a connection; no flow-specific states whatsoever are maintained in hosts and routers along the path between two endpoints. To maintain the robustness of the Internet, a so-called *soft-state* approach is being followed. The reservation state will be cached and periodically refreshed by the end hosts. An unused state will time out after a given period.

Routers
The functional components of an IS-capable router are shown in Figure 67 on page 185. There are two broad functional divisions; the background code (shown with a shaded background) and the forwarding path. As the forwarding path is executed for every IP packet, it must be highly optimized and will probably need hardware assistance. The background code will be loaded into router memory and executed by a general-purpose CPU.

Application Hosts

In order to state its resource requirements, an application will need to specify the desired QOS using a list of parameters called a *flowspec*. The application will need to invoke a local reservation setup agent; how this will be done is not yet specified. In addition, the flowspec will be carried by the reservation setup protocol, passed to admission control in nodes along the setup path, and ultimately used to control the packet scheduling mechanism.

8.2.1.2 Quality-of-Service Requirements

The core service model of IS is concerned mainly with the time-of-delivery of packets. Therefore, the per-packet delay will be the central quantity with which the network makes QOS commitments. The authors go even further and assume that the *only* quantity of importance used to make QOS commitments will be the maximum and minimum delays.

For QOS, applications can be divided into two groups:

Real-Time Applications

These applications need the data in each packet by a certain time; otherwise the data is worthless to the application. Here the average delay is of little importance, since late packets will have an effect on such things as distortion. It is expected that such applications will need to adjust themselves to the service level that the network can provide. This adjustment to using less network resources is paid for by reduced quality. Such applications are called *adaptive*. To do this, admission control must check that the requested flow parameters can actually be accommodated by the network, and the application must understand feedback from the network when it is necessary to reduce the rate of input. Adaptive applications will have the advantage that they can adjust to current network conditions and, therefore, still keep on working.

Elastic Applications

Elastic applications will always wait for data to arrive. For such applications, the average delay will be of importance for application performance. An as-soon-as-possible service model would be suitable for such applications.

8.2.1.3 Resource Sharing Requirements

QOS commitments dictate how the network allocates resources among individual flows. It does not address policy issues for a collection of flows. For resource sharing, it is assumed that the quantity of primary interest is aggregate bandwidth on individual links.

Multiprotocol sharing is needed to prevent one protocol from overloading a link and excluding all other protocols. Multiservice sharing is needed to ensure that real-time traffic does not exclude elastic traffic. Multientity sharing is needed when several departments or companies share a link, and each one desires a specified minimum bandwidth, but the maximum bandwidth may be unspecified to allow full use of a link when it is underutilized by other parties sharing the link. The link-sharing service model is to share the aggregate bandwidth according to some specific shares.

Admission control will be necessary to ensure that link-sharing commitments can be met.

8.2.1.4 Packet Dropping

Up until now it was assumed that all packets within a flow were equally important. However, in audio and video streams some packets are more important than others. It is proposed that a *preemptable* packet service be implemented. Some packets should be marked as preemptable; in cases where a router is in danger of not meeting its service commitments, preemptable packets could be discarded. This has various advantages over just delaying a packet.

8.2.1.5 Reservation Model

This describes how an application negotiates with a network for a specific QOS. The simplest model is where the application requests a QOS and the network either grants it or refuses it. As many applications will be able to get acceptable service from differing QOS levels, some sort of negotiation will be required.

8.2.2 IP Resource Reservation in ATM Networks

The IETF RFC 1932, *IP over ATM: A Framework Document* (see 5 on page 229) discusses various issues in the debate about IP over ATM and is an attempt to focus on issues that must still be resolved. In addition, it is a good level-set for the status of various proposals.

One point for debate is the trade-off between connectionless IP traffic and connection-oriented ATM traffic. At one end of the spectrum, IP would remain highly connectionless and router-to-router traffic would use a single ATM VC. At the other end, a separate ATM VC would be set up for each identifiable IP flow.

ATM offers QOS and traffic management for certain types of services. It may well be advantageous to use ATM VCs where QOS matches the IP flow closely. Many types of IP and TCP traffic are of short duration and the latency involved in setting up an ATM VC would be prohibitive. For such flows it may well be best to just route the traffic as before; ATM VCs can be used as router-to-router connections where they already exist. For other kinds of traffic, it remains to be seen whether a routed solution, such as the Integrated Services Model, will be able to provide the QOS required. The routing overhead, plus cell-to-packet reassembly and disassembly required, would certainly suggest that a direct ATM VC would be a better solution.

RSVP (see 8.3, “Resource Reservation Protocol (RSVP)” on page 189) describes a resource reservation protocol for IP that, while not limited to ATM networks, will be capable of negotiating with a QOS capable link-layer medium (such as ATM) to obtain the QOS requested by an application. RSVP would then make IP applications *ATM aware* without the need for the IP application to communicate directly with a native ATM DLC. For IP applications to make effective use of resource reservation, changes will need to be made to the application.

There are two large differences between how ATM networks work and how RSVP will work that still must be resolved:

- Whereas in ATM connections QOS is static and exists for the duration of a VC, RSVP allows for the changing of QOS on-the-fly. It remains to be seen how RSVP will handle this when used with ATM networks.
- ATM networks are source oriented, which means that an ATM sender is responsible for establishing a point-to-multipoint VC (although UNI 4.0 will implement a *leaf initiated join* function). Leaves on that VC must send messages to the source telling it to remove them from the multipoint tree. RSVP, on the other hand, expects a receiver to send reservation requests back up the path to the sender, and not directly back to the sender. Also, as an ATM point-to-multipoint VC is uni-directional, some other mechanism must be found to route RSVP flows back to the sender.

8.3 Resource Reservation Protocol (RSVP)

The Internet Draft, draft-ietf-rsvp-spec-10 (see 13 on page 230), describes version 1 of RSVP. This section provides an overview of the functions of RSVP, based on the Introduction of the Version 1.0 edition of the Internet Draft.

RSVP is used by a host, on behalf of an application data stream, to request a specific quality of service (QOS) from the network. The RSVP protocol is also used by routers to deliver QOS requests to all nodes along the path(s) of the data stream and to establish and maintain state to provide the requested service. RSVP requests will generally, although not necessarily, result in resources being reserved along the data path.

RSVP requests resources for *simplex* data streams; that is, it requests resources in only one direction. Therefore, RSVP treats a sender as logically distinct from a receiver, although the same application process may act as both a sender and a receiver at the same time. RSVP operates on top of IP (either IPv4 or IPv6), occupying the place of a transport protocol in the protocol stack. However, RSVP does not transport application data but is rather an Internet control protocol, such as ICMP, IGMP, or routing protocols. Like the implementations of routing and management protocols, an implementation of RSVP will typically execute in the background, not in the data forwarding path.

RSVP is not itself a routing protocol; RSVP is designed to operate with current and future unicast and multicast routing protocols. An RSVP daemon consults the local routing database(s) to obtain routes. In the multicast case, for example, a host sends IGMP messages to join a multicast group and then sends RSVP messages to reserve resources along the delivery path(s) to that group. Routing protocols determine where packets get forwarded; RSVP is only concerned with the QOS of those packets that are forwarded in accordance with routing.

Each node that is capable of resource reservation passes incoming data packets through a *packet classifier*, which determines the route and QOS class for each packet. For each outgoing interface, a *packet scheduler* then makes forwarding decisions for each packet to achieve the promised QOS on the particular link-layer medium used by that interface.

If the link-layer medium is QOS-active, that is, if it has its own QOS management capability, then the packet scheduler is responsible for negotiation with the link layer to obtain the QOS requested by the RSVP. This mapping to the link-layer QOS may be accomplished in a number of possible ways; the details will be media-dependent. On a QOS-passive medium such as a leased line, the scheduler itself allocates packet transmission capacity. The scheduler may also allocate other system resources, such as CPU time or buffers.

In order to efficiently accommodate heterogeneous receivers and dynamic group membership, RSVP makes receivers responsible for requesting QOS. A QOS request, which typically originates from a receiver host application, is passed to the local RSVP implementation. The RSVP protocol then carries the request to all the nodes (routers and hosts) along the reverse data path(s) to the data source(s).

At each node, the RSVP daemon communicates with two local decision modules, *admission control* and *policy control.* Admission control determines whether the node has sufficient available resources to supply the requested QOS. Policy control determines whether the user has administrative permission to make the reservation. If both checks succeed, the RSVP daemon sets parameters in the packet classifier and scheduler to obtain the desired QOS. If either check fails, the RSVP program returns an error notification to the application process that originated the request. We refer to the packet classifier, packet scheduler, and admission control components as *traffic control.*

RSVP is designed to scale well for very large multicast groups. Since both the membership of a large group and the topology of large multicast trees are likely to change with time, the RSVP design assumes that a router's state of traffic control will be built and destroyed incrementally. For this purpose, RSVP uses *soft state* in the routers. That is, RSVP sends periodic refresh messages to maintain the state along the reserved path(s); in absence of refreshes, the state will automatically time out and be deleted.

RSVP protocol mechanisms provide a general facility for creating and maintaining a distributed reservation state across a mesh of multicast and unicast delivery paths. RSVP transfers reservation parameters as opaque data (except for certain well-defined operations on the data), which it simply passes to traffic control for interpretation. Although the RSVP protocol mechanisms are largely dependent on the encoding of these parameters, the encoding must be defined in the reservation model that is presented to an application.

In summary, RSVP has the following attributes:

- RSVP makes resource reservations for both unicast and multicast and many-to-many multicast applications, adapting dynamically to changing group membership as well as changing routes.
- RSVP is simplex, that is, it reserves for a data flow in one direction only.
- RSVP is receiver-oriented, that is, the receiver of the data flow initiates and maintains the resource reservation used for that flow.
- RSVP maintains *soft state* in the routers, providing graceful support for dynamic membership changes and automatic adaptation to routing changes.
- RSVP provides several reservation models or *styles* to fit a variety of applications.
- RSVP provides transparent operation through routers that do not support it.

8.3.1 Resource Reservation

This section gives a brief overview of how RSVP works.

8.3.1.1 RSVP Messages

Central to the functions of RSVP are two types of information:

Path Messages

Sending nodes that implement RSVP periodically send *path* messages into the network; these are forwarded by routers downstream along the unicast or multicast routes provided by existing routing tables. The path messages store the path state in each node along the way. This includes at least the unicast IP address of the previous hop node. The path message also contains the sender's *flowspec*.

Flowspec

The flowspec may contain the sender's *Template*, *TSpec* and *Adspec*.

Sender Template

The template has the form of a filter specification that describes the format of data packets that the sender will generate. This is used to differentiate between this sender's packets and other packets in the same session on the same link. RSVP defines a *session* as a data flow with a particular destination address and transport-layer protocol, where the destination address could be the IP destination address and a *generalized destination port*; at this time, only a TCP/UDP port number is supported as a generalized port.

Filter specs and sender templates specify the pair {SrcAddress, SrcPort}, where SrcPort is the UDP/TCP source port field. Because the UDP/TCP port numbers are used for traffic classification, intermediate routers must be able to examine these fields in the packets they route.

TSpec

The TSpec contains the traffic characteristics of the data the sender will transmit.

Adspec

An Adspec received in a path message is passed to local traffic control, which returns an updated Adspec. The updated Adspec is then forwarded in path messages sent downstream.

RESV Messages

The RESV message contains the actual reservation request. RESV messages are sent hop-by-hop back along the path to the unicast address of the previous hop, the address being obtained from the path message.

8.3.1.2 RSVP Flows

As the receiver makes resource reservations in RSVP, the receiver must first be informed about what reservations need to be made. The information used is extracted from the path messages. Each router that path messages pass through, and the ultimate receiver of the messages, sends RESV messages back to the previous hop once the requested reservations have been made. If the reservation cannot be made, then an error message is returned back up the path to the sending node. The receiver sends an RESV message periodically as long as it needs to retain a reservation. The intermediate nodes would remove the reservation after a time-out period if this periodic RESV message was not received. If a router or link fails, and the data is rerouted around the failure, then the path messages (which are also sent periodically) will be sent over the new route. This will trigger RESV messages along the new route that causes the appropriate reservations to be made. Reservations that were in place along the original route time out after a given period and free up resources again.

The reservations in place can be dynamically updated. To change them, the sender only needs to start sending revised path messages. If a received state differs from a stored state, then the stored state is updated. If a state update has occurred, then RESV messages are returned to propagate these state changes end-to-end.

Appendix A. Protocol Stack Reference

A.1 ATM Layers

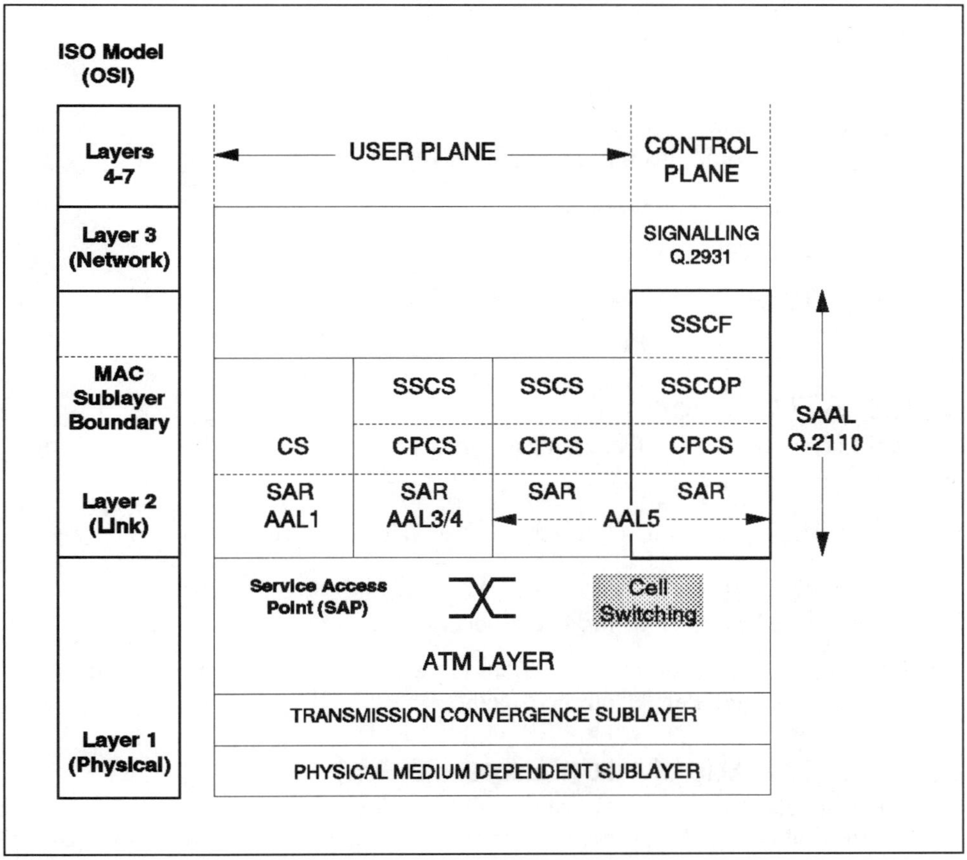

Figure 68. ATM Layers

Figure 68 shows a diagram of the ATM interface broken down into layers, shown next to the OSI model. Refer to Figure 68 when reading the following descriptions.

A.1.1 Physical Layer

The physical layer contains a Physical Medium Dependent (PMD) sublayer that performs media-specific functions and a Transmission Convergence sublayer (TC) that performs functions common across different physical media.

Physical Medium Dependent Sublayer (PMD)

The PMD provides the following functions:

- Encoding for transmission
- Timing and synchronization
- Transmission and receival (electrical or optical)

Transmission Convergence Sublayer (TC)

The TC provides the following functions:

- HEC generation and checking
- Cell delineation
- Transmission frame adaptation
- Decoupling of cell rate

A.1.2 ATM Layer

The ATM layer provides the forwarding and multiplexing functions. These include the following:

- Flow control and queuing priority to meet service guarantees for delay
- Cell loss priority
- Payload type identification (user cells versus control or management cells)
- Multiplexing and demultiplexing of the cells associated with different virtual paths and virtual channels on the physical media.

A.1.3 ATM Adaptation Layer (AAL)

The ATM Adaptation Layer (AAL) can be split into three sublayers. The common part convergence sublayer (CPCS) and segmentation and reassembly sublayers (SAR) are common to both control planes and user planes.

Common Part Convergence Sublayer (CPCS)

The Common Part Convergence Sublayer (CPCS) is defined in ITU recommendation I.363. The CPCS is a sublayer of the Convergence Sublayer of the SAAL. Functions performed by the CPCS are:

- End-to-end transport of service data units

- Error detection and handling
- Provide information on buffer allocation requirements
- Provide sequence integrity for service data units

Segmentation and Reassembly Layer (SAR)

The SAR provides the services whereby SAR service data units (SDUs) are segmented to fit outgoing ATM cells, while incoming cells are reassembled into SDUs and passed up to the CPCS. Figure 36 on page 94 shows a basic representation of this.

A.1.3.1 User Plane

Service Specific Convergence Sublayer (SSCS)

The Service Specific Convergence Sublayer (SSCS) perfoms functions specific to the data service class that is being supported by the AAL. The SSCS builds on the functions provided by the CPCS and SAR.

A.1.3.2 Control Plane

Signalling ATM Adaptation Layer (SAAL)

The Signalling ATM Adaptation Layer (SAAL) consists of a Segmentation and Reassembly (SAR) function and a Convergence sublayer. The SAAL comprises AAL functions necessary to support a signalling entity. The SAAL makes use of the service provided by the CPCS and the SAR which form the common part of AAL type 5. It is capable of providing an assured data transfer, thereby minimizing the chance that lost or corrupted information might cause a signalling procedure to fail. The complete structure and specification of the SAAL is defined in recommendation ITU Q.2100.

ITU Q.2931

This recommendation specifies the procedures for the establishing, maintaining and clearing of network connections at the B-ISDN UNI. It is defined as an OSI layer 3 procedure that requests services from SSCOP/AAL5 through primitives defined by the Service Specific Coordination Function (SSCF). Q.2931 uses an dedicated out-of-band channel (VPI=0, VCI=5) to communicate with the ATM network. This channel is statically defined and always connected to the network.

The following capabilities are supported by Q.2931:

- Request point-to-point and point-to-multipoint connections
- Request connections with:
  - Symmetric or asymmetric bandwidth requirements
  - Different QOS attributes
- Perform end-to-end compatibility exchange and negotiation

Q.2931 is used by a resource manager, application or protocol stack that needs to establish an ATM connection.

Service Specific Coordination Function (SSCF)

The Service Specific Coordination Function (SSCF) maps the service of the Service Specific Connection-Oriented Protocol (SSCOP) of the AAL to the needs of layer 3 protocols for access signalling across the UNI and NNI. Figure 68 on page 193 shows where the SSCF resides in the SAAL layer. This structure allows a common connection-oriented protocol to provide a generic reliable data transfer service for different AAL interfaces defined by the SSCF. To date, a UNI SSCF and a NNI SSCF have been defined. Recommendation ITU Q.2130 covers the function of the SSCF within the AAL structure at the UNI and Recommendation ITU Q.2140 covers the SSCF at the NNI.

Service Specific Connection Oriented Protocol (SSCOP)

The Service Specific Connection Oriented Protocol (SSCOP) provides assured data delivery between AAL connection endpoints. The SSCOP is part of the Service Specific Convergence Sublayer (SSCS) which is a sublayer of the Signalling ATM Adaptation Layer (SAAL). The SSCOP is a peer-to-peer protocol which provides the following functions:

- Transfer of user data with sequence integrity
- Error correction by selective retransmission
- Flow control
- Connection control
- Error reporting to layer management
- Connection maintenance in the prolonged absence of data transfer
- Local retrieval by the user
- Error detection of protocol control information
- Status reporting

The SSCOP is described in detail in recommendation ITU Q.2130 and recommendation ITU Q.2110.

A.2 SNA Layers

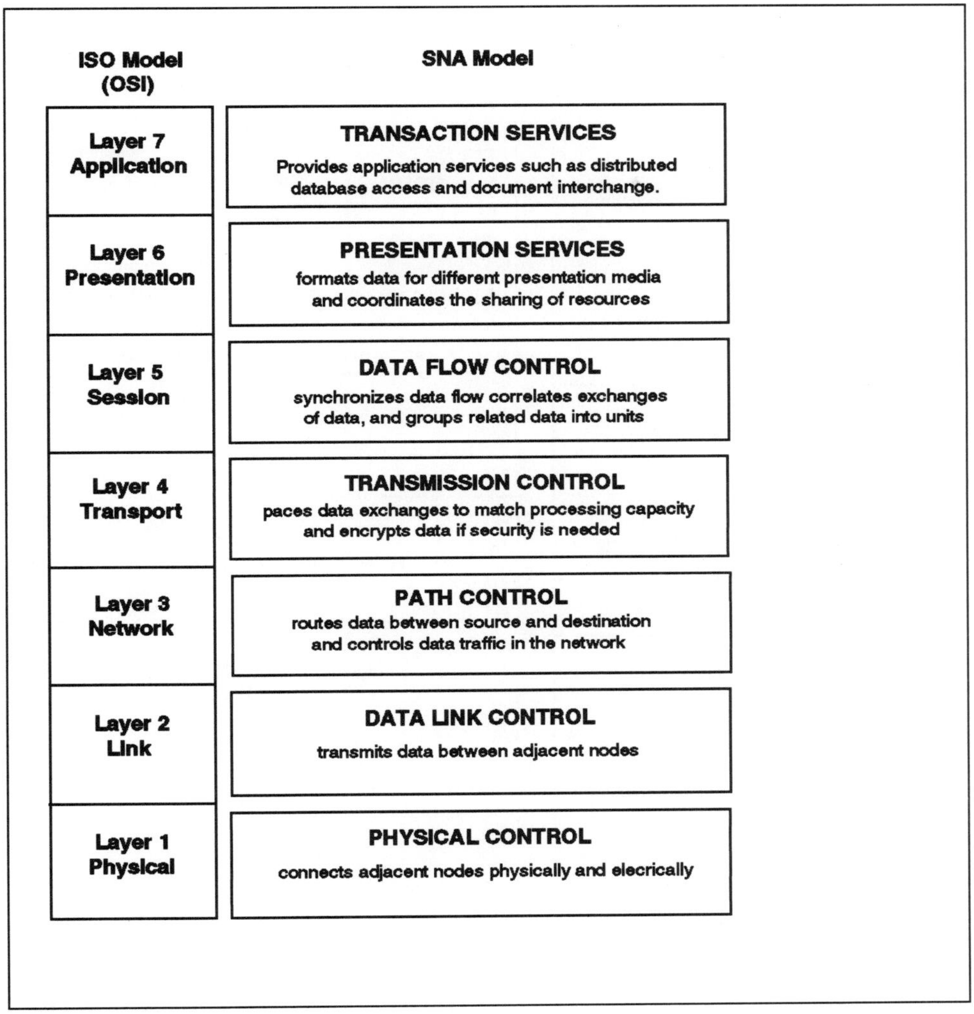

Figure 69. SNA Layers

A.3 TCP/IP Layers

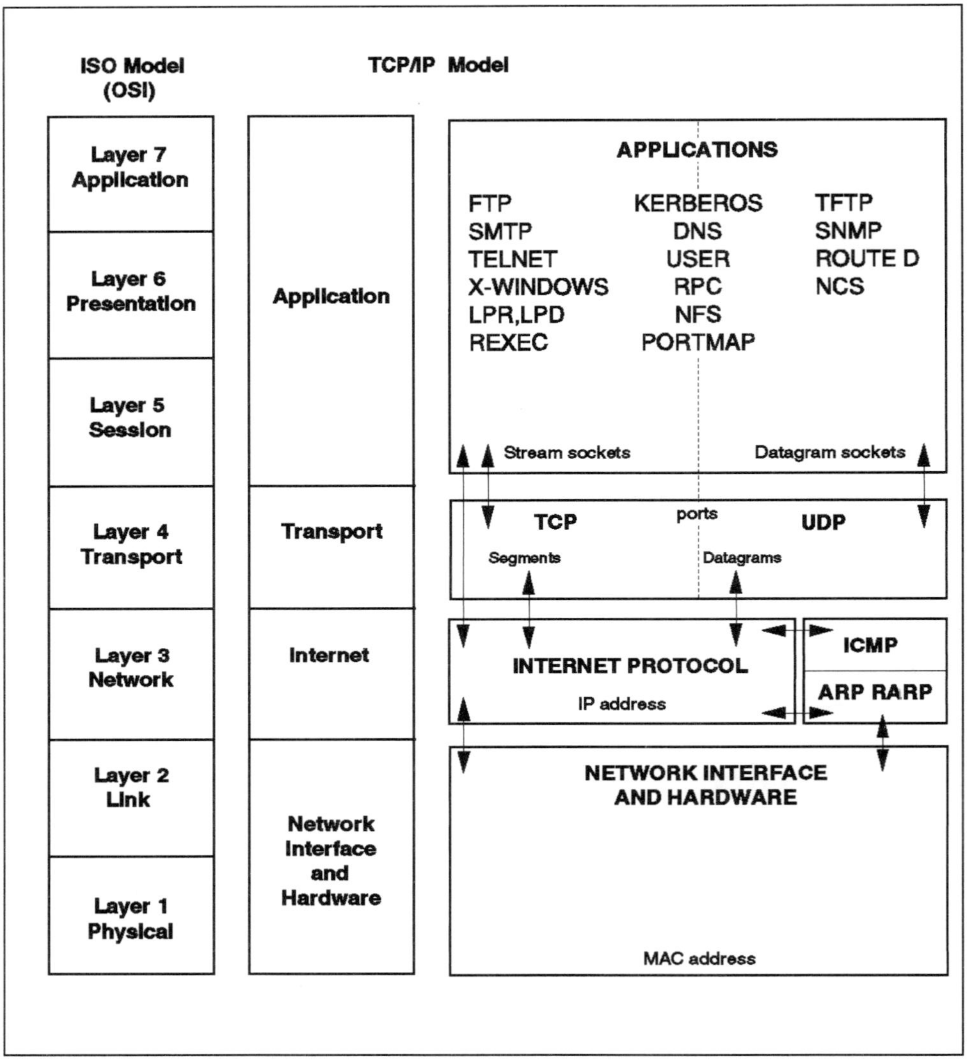

Figure 70. TCP/IP Layers

Appendix B. ATM Service Categories

Traffic management is probably the most controversial (and important) aspect of ATM. The original concept of ATM included the idea that network congestion would not be controlled by mechanisms within the network itself. It was (and still is, by many) felt that detailed flow controls cannot be successfully performed within the network. The ATM Forum has specified five "service categories" in relation to traffic management in an ATM network. These are similar to but are not the same as the AAL service classes. These categories are:

Constant Bit Rate (CBR)

CBR traffic includes anything where a continuous stream of bits at a predefined constant rate is transported through the network. The network must reserve the full bandwidth requested by the peak cell rate (PCR) specified when the connection is set up.

This might be voice (compressed or not), circuit emulation (say the transport, unchanged, of a T1 or E1 circuit), or some kind of video, all data streams that are very sensitive to network delay. Typically you need both short transit delay and very low jitter in this service class.

Real-Time Variable Bit Rate (rt-VBR)

This is like CBR in the sense that we still want low transit delay but the traffic will vary in its data rate. We still require a guaranteed delivery service. The data here might be compressed video, compressed voice with silence suppression, or HDLC link emulation with idle removal.

Non-Real-Time Variable Bit Rate (nrt-VBR)

This is again a guaranteed delivery service where transit delay and jitter are perhaps less important than in the rt-VBR case. An example here might be MPEG-2 encoded video distribution. In this case, the information may be retrieved from a disk and be one-way TV distribution. A network transit delay of even a few seconds is not a problem here. But we do want guaranteed service because the loss of a cell in compressed video has quite a severe effect on the quality of the connection.

Unspecified Bit Rate (UBR)

The UBR service is for "best-effort" delivery of data. It is also a way of allowing for proprietary internal network controls. A switch using its own (nonstandard) internal flow controls should offer the service as UBR class. You send data on a UBR connection into the network and if there is any congestion in any resource, then the network will throw your data away. In many cases, with appropriate end-to-end error recovery protocols this may be quite acceptable. This should be workable for many if not most traditional data applications such as LAN emulation and IP transport.

Available Bit Rate (ABR)

The concept of ABR is to offer a guaranteed delivery service (with minimal cell loss) to users who can tolerate a widely varying throughput rate. Like UBR, ABR uses any excess network bandwidth but also exploits traffic management to avoid cell loss if network congestion occurs. The idea is to use whatever bandwidth is available in the running network after other traffic utilizing guaranteed bandwidth services has been serviced. ABR though does need a guaranteed minimum cell rate to keep applications that use ABR running, and not allow CBR and VBR traffic to consume all the network resources and leave no available bandwidth for ABR. One statement 1 on page 229 of the primary goal of the ABR service is for *"the economical support of applications with vague requirements for throughputs and delays"*.

In an operational network, there may be bandwidth "allocated" to a particular user but in fact going unused at this particular instant in time.

Either by providing feedback from the network to the sender or by monitoring the network's behavior, the ABR service can change the bit rate of the connection dynamically as network conditions change. The end-user system must be able to obey the ABR protocol and to modify its sending rate accordingly.

The ABR specification requires a closed-loop congestion control mechanism. However there are several levels of feedback information that can be implemented.

Explicit Forward Congestion Indication (EFCI)

This is the simplest method and uses and end-to-end control loop. Switches produce EFCI messages that are piggybacked onto data cells; these notify the destination station of the congestion. The destination then sends a resource management cell back to the traffic source station telling it to reduce its transmission rate.

With EFCI the switches have no closed loop feedback mechanism, the loop is closed by the end station. The end stations are responsible for flow control.

Explicit Rate Marking (ERM)

ERM provides a slightly enhanced feedback mechanism in that a switch can alter the feedback information being passed back through the network to the source station. The resource management cells are "marked" as they flow back through the switches to indicate when larger traffic reductions are needed than specified by the destination station.

While ERM and EFCI switches can coexist, the results are only acceptable when all switches support ERM.

Segmented Virtual Source/Virtual Destination (VS/VD)

A step further than EFCI and ERM, VS/VD capable switches can themselves respond to EFCI notification just like the end station would. They act as a "virtual end stations" and can send their own resource management cells back to the source station. VS/VD speeds up the reaction of feedback loops by reducing the delay before congestion is reported back along a now smaller feedback loop.

A VS/VD-capable switch must contain a degree of intelligence that will add its cost, but not every switch in a network needs to be VS/VD-capable. Strategically placed VS/VD switches can provide acceptable control. Placing VS/VD switches at the interface between a LAN and a WAN can isolate the WAN from LAN stations that transmit into the WAN at speeds far greater than the WAN can handle (for example, non-ABR compliant end stations or stations that ignore resource management cells with feedback information.

Hop-by-Hop Virtual Source/Virtual Destination (VS/VD)

Hop-by-Hop VS/VD takes the segmented model one step further and has only VS/VD capable switches. Feedback occurs at ever hop in the network providing the greatest possible control and smallest feedback loops.

For ABR to function without cell loss, switches will need to have enough buffers available to buffer incoming ABR traffic while waiting for congestion control to take effect. In addition, each virtual connection must have its own buffer queue to prevent well-behaved end stations from being affected by end stations causing congestion. If buffers were shared, a misbehaving end station will fill all available buffers for a VC before congestion control kicks in. There would be no buffer space left to provide cell-loss buffering for a well-behaved end station, its cells would then be discarded even though it was adhering to its traffic contract.

Appendix C. AAL Service Classes

The ATM adaptation Layer provides the functions associated with different service classes defined in ATM. The following service classes are defined:

Class A

Connection-oriented, constant bit rate (CBR) service with timing required between the source and destination (for example, circuit emulation). Uses AAL type 1.

Class B

Connection-oriented, variable bit rate (VBR) service with timing required between the source and destination. Uses AAL type 2 which is currently undefined. Work is underway within the ITU-T evaluating the needs and requirements for this AAL type.

Classes C & D

Connection-oriented or connectionless variable-bit-rate service without timing required between the source and destination. Uses AAL type 3/4 or AAL type 5. AAL3/4 is a combined AAL for the connection and connectionless service classes, but its complexity has led to the creation of AAL5, sometimes called Simple and Efficient Adaptation Layer (SEAL). AAL5 is commonly used for data communication but is also being used for video delivery when the time base recovery is performed by the user.

AAL5 permits the use of a non-null SSCS:

- The Frame Relay Service Specific Convergence Sublayer (FR-SSCS) provides a frame relay emulation service.
- The Service Specific Connection-Oriented Protocol (SSCOP) SSCS is used to provide reliable data delivery. It was initially defined to support the ATM UNI signalling protocol (Q.2931). AAL5 with the SSCOP is referred to as the Signalling ATM Adaptation Layer (SAAL).

Class X

Class X is a connection-oriented ATM Transport where the AAL, traffic type (VBR or CBR), and timing requirements are user-defined and transparent to the network. The user chooses only the desired bandwidth and QOS at the time the connection is set up to establish a Class X connection.

Appendix D. ATM Address Formats

The address formats used in ATM are shown in Figure 71 on page 206. There are three different address formats used here:

ITU-T (E.164) Format

This format is essentially a telephone number style address. It is specified by the ITU-T and will be used by public (carrier provided) ATM networks.

DCC (Data Country Code) Format

This format carries LAN addresses as specified by the IEEE 802 recommendations.

ICD Format

This is the format specified by the International Organization for Standardization (ISO) for Open Systems Interconnection (OSI).

The ATM Forum specifies that equipment must support all three formats in private networks. Networks supporting the ITU-T specification (mainly public ATM networks) only need to support the ITU-T address.

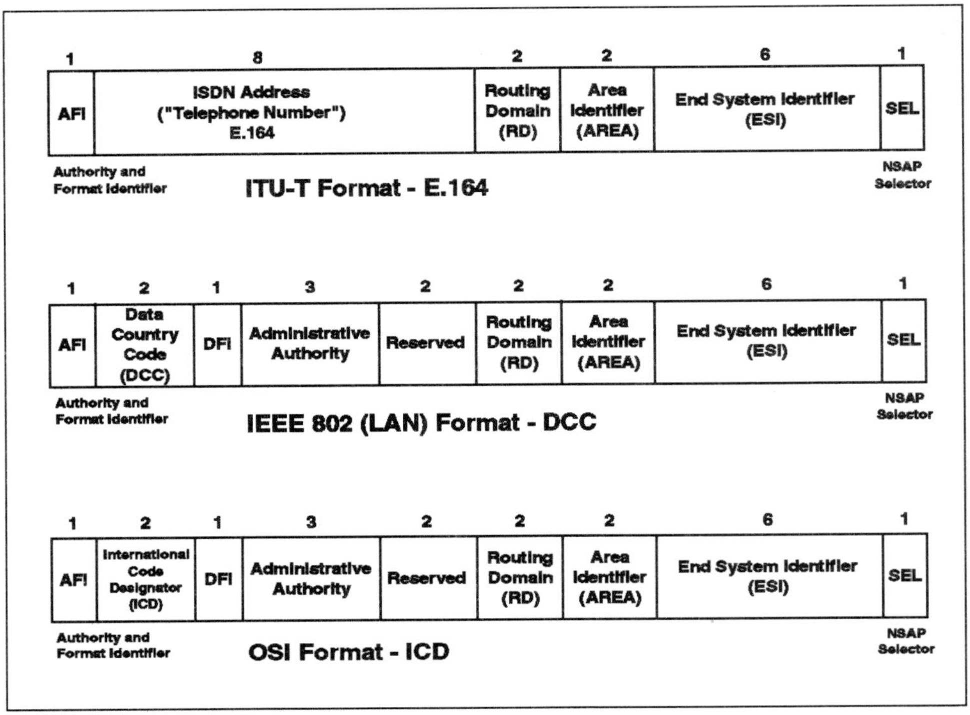

Figure 71. *ATM Address Formats*

Some people have objected to the use of 20-byte addresses. Twenty bytes is 160 bits or about 10^{40} possible addresses. This is sufficient address range to uniquely number every cell in every human body on earth! This is said to be wasteful.

The reason the address is so long is because it is structured. Structured addresses greatly simplify the task of locating particular endpoints and, in any case, the addresses are only carried in the setup request and so the overhead is minimal.

Appendix E. Multiprotocol Encapsulation over AAL 5 (RFC 1483)

RFC 1483 describes two encapsulation methods for carrying connectionless network interconnect traffic, routed and bridged protocol data units (PDUs) over an ATM network:

LLC Encapsulation
: Enables multiplexing of multiple protocols over a single ATM virtual circuit

VC-Based Multiplexing
: Where each protocol is carried over a seperate ATM virtual circuit

Full details of RFC 1483 can be found in 7 on page 229. The RFC 1483 mechanism is used with RFC 1577, MPOA, I-PNNI and APPN/HPR over ATM and is hence an overview is included here for completeness.

E.1 LLC Encapsulation

LLC encapsulation saves on PVC/SVC setup and hence is desirable when it is not practical to have a seperate VC for each protocol, for example in a public network where the tarrif charge is based on the number of VCs.

LLC Encapsulation for Routed Protocols

In LLC encapsulation each packet is prefixed by the associated IEEE 802.2 LLC header, which is possibly followed by an IEEE 802.1 a Subnetwork Attachment Point (SNAP) header. This prefix identifies the protocol of the routed packet.

LLC Encapsulation for Bridged Protocols

In LLC encapsulation bridged packets are identified by the bridged media type in the SNAP header.

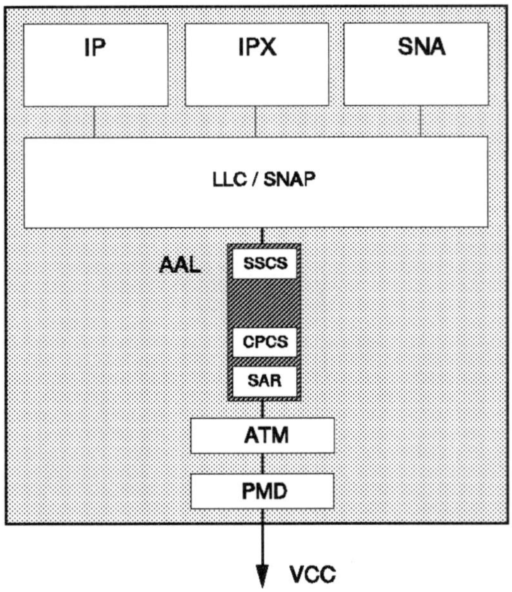

Figure 72. LLC Encapsulation

E.2 VC-Based Multiplexing

In VC-based multiplexing, the carried network interconnect protocol is identified implicitly by the VC connecting the two ATM stations, that is, each protocol must be carried over a seperate VC. There is therefore no need to include explicit multiplexing information in the packet. Hence, VC-based multiplexing (null encapsulation) is the simplest approach and provides a greater granularity of control, although more costly in terms of PVC/SVC usage.

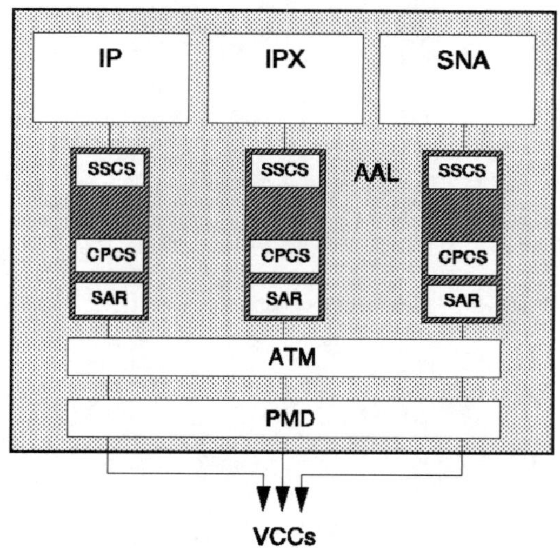

Figure 73. *VC-Based Multiplexing*

Appendix F. Cells in Frames (CIF)

While the main emphasis of this book has been on transporting various protocols over ATM networks, a new proposal for transporting ATM cells over Ethernet V2 networks is also worth mentioning here. The information in this appendix is taken from the *Cells In Frames Version 1.0* draft 24 on page 231.

The above document specifies the mechanisms by which ATM traffic is carried across a media segment and network interface card conforming to the Ethernet Version 2 specification. The mechanisms are collectively referred to as *Cells In Frames* (CIF). ATM cells can be carried over many different physical media, from optical fiber to spread spectrum radio. ATM is not coupled to any particular physical layer. CIF defines a new physical layer over which ATM traffic can be carried. It is not simply a mechanism for translation between frames and cells; neither is it simple encapsulation.

While the concept of carrying cells in legacy LAN frames is not new, the methods proposed for CIF have some unique features that enhance performance and adaptability. The combination of CIF end system software and a CIF attachment device (CIF-AD) will make it possible to support ATM service, including multiple classes of service, over an existing Ethernet NIC, just as if a specialized ATM NIC were in use. This document specifies how the ATM layer protocols can be made to work over the existing Ethernet framing protocol in such a way that operation is transparent to an application written to an ATM-compliant API.

An ATM end system will possess an ATM signalling module, an ILMI module and the ability to process multiple AAL types. Together these are the *ATM processing layer*. In order to provide a generic solution that works with the existing Ethernet driver on the CIF end system, a *shim* will bridge between the ATM processing layer and the existing Ethernet driver. Legacy applications will have access to the AAL layers via a LAN Emulation module, MPOA or classical IP over ATM, just as with other end system ATM implementations.

The CIF-AD receives groups of cells from the end system and forwards them to an ATM backbone switching fabric. The CIF-AD might be implemented as a multiplexor or, more probably, an ATM switch. For CIF functionality, the CIF-AD does not need to implement LANE, MPOA, or Classic IP. Figure 74 on page 212 shows end systems attached via point-to-point 10BASE-T wiring to the CIF-AD. This is the expected configuration.

One of the goals of the mechanisms described in this specification is to ensure that the CIF end system is not unduly burdened by ATM processing requirements. Thus, some of the ATM processing is moved from the end system to the attachment device. There are three ATM functions that would impair performance if they were implemented in software in an end system: ABR traffic management (especially RM cell management), AAL5 CRC calculation, and the actual generation of cells by the ATM layer.

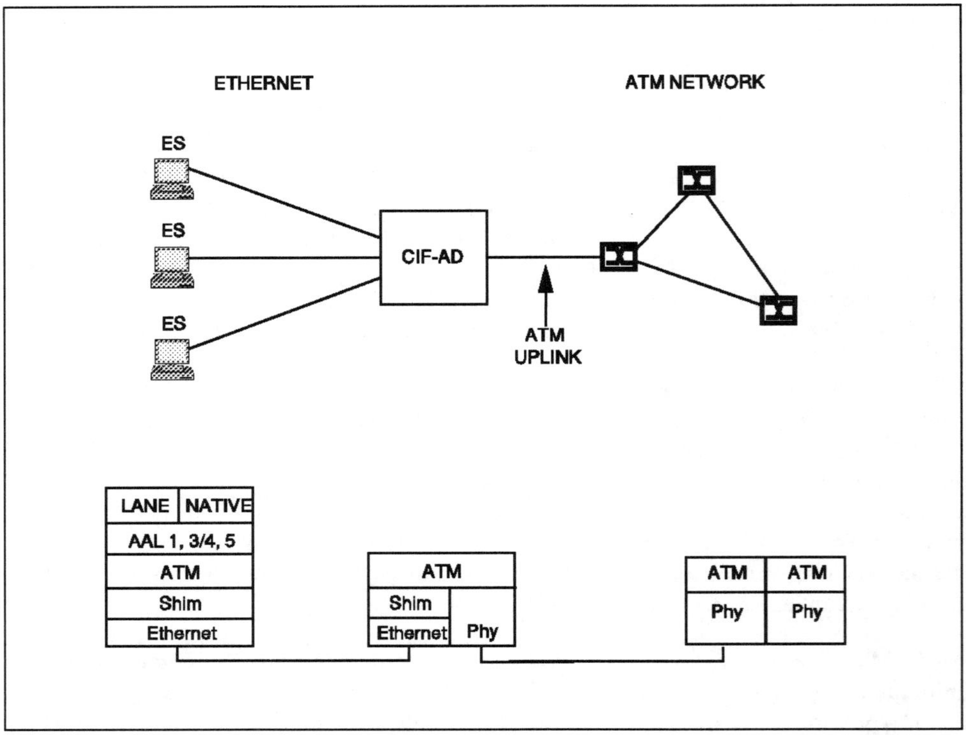

Figure 74. Cells In Frames Protocols Stacks

This division of labor is essentially at the ATM SAP, the boundary between the AAL layer and the ATM layer. ATM cell payloads are packed in Ethernet frames along with parameters which describe the contents of the payloads' cell headers. In Ethernet frames sent from the end system to the attachment device, parameters are specified for cell headers, which should be constructed for those cell payloads by the attachment device. In Ethernet frames sent from the CIF-AD to the end system, the parameters describe the contents of the cell headers that were received with the payloads.

CIF includes:

- Specification of the framing for carrying cell payloads on the Ethernet
- The generation and interpretation of the frames and action to be taken based on the contents of the frames
- Control
- Management

Perhaps the best way to describe CIF is to imagine it as a *Distributed* SAR layer, one part being in the end system and the other part in the CIF-AD, with Ethernet as the protocol used to connect both parts of the distributed SAR function (see Figure 76 on page 216).

F.1.1 Framing

On an Ethernet, CIF frames shall have a standard Ethernet Version 2 header and trailer. CIF devices may be configured to support 802.3 as well, but there is no mechanism defined in CIF for negotiating framing. If the CIF-AD and end systems support 802.3, it can be configured manually. Refer to Figure 75 on page 214 for the headers described in the following descriptions.

Note: CIF Ethernet frame headers shall use a new Ethernet type, which is not yet assigned.

The first 4 bytes of the Ethernet payload shall contain the CIF header. The CIF header contains the information that would ordinarily be passed as parameters from the AAL layer to the ATM layer, across the ATM service access point. The structure and semantics of the fields in the CIF header are the same as those of the first 4 bytes of a UNI cell header, so that it can be used by the CIF-AD as a cell.

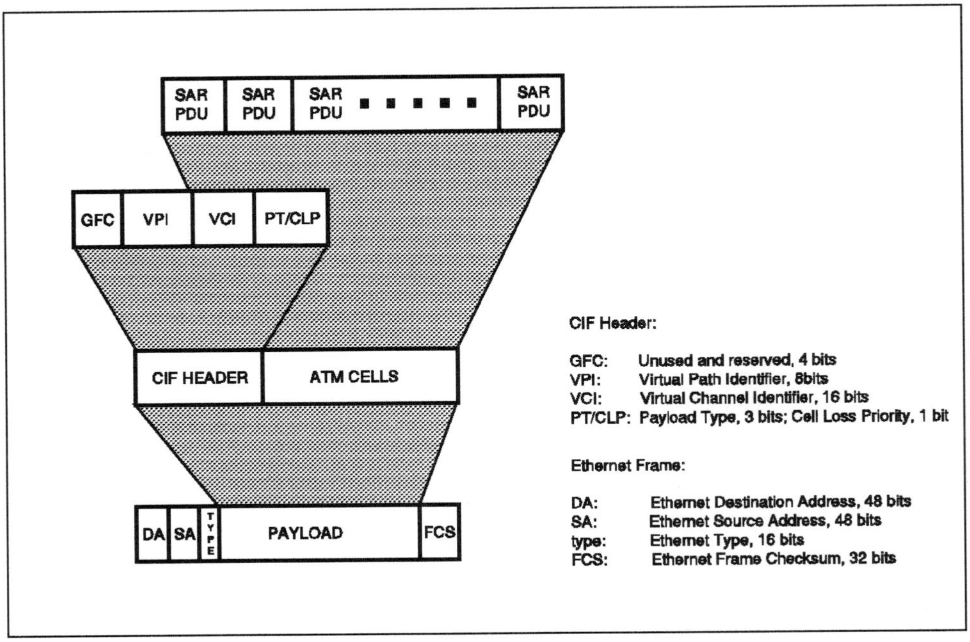

Figure 75. *CIF and Ethernet Headers*

The CIF header is followed by up to 31, 48-byte cell payloads (SAR-PDUs), or 1488 bytes, contiguously without cell headers. All of the SAR-PDUs within a single Ethernet frame shall be destined to the same VC. SAR-PDUs shall be complete (for example, the AAL1 SAR-PDU header byte shall be calculated and inserted) and a frame shall contain no partial SAR-PDUs.

The minimum CIF frame, containing one cell, consists of 14 bytes of Ethernet header, 4 bytes of CIF header, 48 bytes of cell payload (SAR-PDU, ATM-SDU), and 4 bytes of Ethernet trailer, for a total of 70 bytes.

All fields in the CIF header are used as defined in the specification for the ATM cell header with the following exceptions:

- The GFC field is reserved and should always be 0.
- Special consideration is required if the SDU-type (in the Payload Type field) is 1.
  - When the end system receives an Ethernet frame from the CIF-AD in which the CIF header SDU-type is 1, the end system shall consider the SAR-PDUs in the frame to all have had an SDU-type of 0 except for the last, which shall be considered to have had an SDU-type of 1.

- When the CIF-AD receives a CIF Ethernet frame in which the CIF header SDU-type is 1, the CIF-AD shall set the SDU-type for all cells generated from the SAR-PDUs in that frame to 0, except the last one, where it shall be set to 1.
- When the CIF-AD is generating a CIF Ethernet frame from cells received on another interface, and adds a SAR-PDU to it from a cell whose SDU-type was 1, it shall set the SDU-type in the frame's CIF header to be 1 and add no more SAR- PDUs to the Ethernet frame.

These rules are true for all current AALs, and should be expected to hold true for future AALs. Only the last SAR-PDU in a frame can be for a cell where the SDU type is 1. In VCs using AAL5 all of the SAR-PDUs in a frame shall belong to the same CPCS-PDU, and in AAL1 a frame where the CIF header SDU type is 1 shall only contain one SAR-PDU.

F.1.2 Generation and Processing of Frames

For traffic from the CIF end system to the network, processing of the AAL SDU starts in the end system and is completed at the CIF-AD where the individual ATM cells are created (see Figure 76 on page 216).

On the send side, the AAL SDU is the basic unit. CPCS processing is peformed, adding any header or trailer, and the resulting CPCS-PDU is broken up into 48-byte payloads at the SAR layer. A CIF header and then an Ethernet header are added to up to 31 of the 48-byte SAR-PDUs (the most that will fit in a single Ethernet frame), and an Ethernet trailer is appended to the end of the set. The Ethernet frame is then sent to the CIF-AD. This process is repeated as necessary until the entire CPCS-PDU has been transmitted. In the CIF-AD, the Ethernet header and trailer for each Ethernet frame are discarded and the CIF header is used to construct an ATM Cell header for each of the ATM cell payloads.

For traffic from the network to the end system, the CIF-AD places one or more cell payloads in an Ethernet frame with an appropriate CIF header. The CIF-AD stops filling a particular frame whenever:

- The number of SAR-PDUs placed in the frame reaches a predetermined maximum for that VC
- The SAR-PDU from a cell with an SDU-type of 1 is placed in the frame

The end system performs the above operations in reverse.

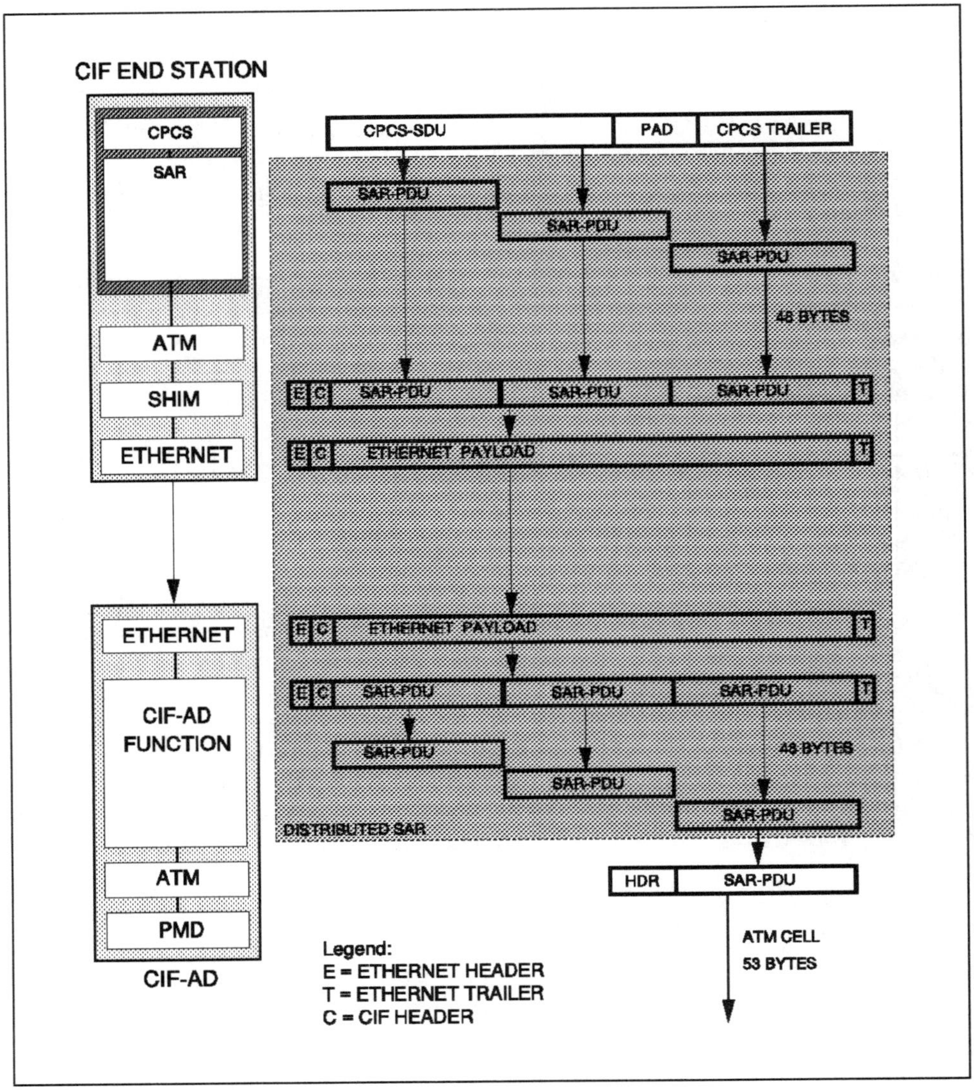

Figure 76. Cells to Frames

F.1.3 ATM Adapation Layer 5 Traffic

The following functions have been defined for AAL5 traffic.

F.1.3.1 End-System to CIF-AD Traffic

The end system shall provide a complete CPCS-PDU to the CIF-AD in one or more Ethernet frames. The end system shall add the appropriate amount of padding and an AAL5 trailer to the data passed to the CIF-AD. The trailer length field and UUI shall be filled in correctly. The first 4 bits of the CRC-32 field shall be filled with the PDU sequence number (see F.1.3.3, "The PDU Sequence Number") The rest of the CRC-32 field shall be ignored by the CIF-AD.

The end system shall segment an AAL5 CPCS-PDU into frames of, at most, 31 SAR-PDUs. Each frame shall begin with a CIF header. If the frame contains the last cell of the packet, then the SDU-type in the CIF header shall be 1.

The CIF-AD shall verify the PDU sequence number as described in F.1.3.3, "The PDU Sequence Number." The CIF-AD shall calculate the AAL5 CRC-32 and insert it into the user's AAL5 CPCS PDU. The CIF-AD shall check the AAL5 length field, and if invalid, insert an invalid CRC-32.

F.1.3.2 CIF-AD to End-System Traffic

The CIF-AD shall calculate and check the CRC for an AAL5 CPCS-PDU arriving from the network. In the last frame of a PDU being sent to the end system, the CIF-AD shall replace the first 4 bits of the CRC-32 field with the PDU sequence number as described in F.1.3.3, "The PDU Sequence Number." The CIF-AD shall replace the rest of the CRC field with all zeros to indicate that a valid CRC was received, and by all ones to indicate validation failure.

The end system shall validate the AAL5 CPCS-PDU by checking the PDU sequence number, CRC-32 validity indication, and length field. The length must be a value greater than or equal to the actual received length minus 40 and less than or equal to received length minus 8 (because of variable padding to fill out the final cell). The PDU shall be discarded if it fails any of the three tests.

F.1.3.3 The PDU Sequence Number

The first 4 bits of the AAL5 CPCS-PDU CRC-32 field are used as an unsigned 4-bit PDU sequence number. When a VC becomes active, the first PDU sent by either the end system or the CIF-AD shall have a sequence number of 1. In subsequent PDUs the sequence number shall be incremented by 1 in each PDU sent. Sequence number 15 shall be followed directly by sequence number 0.

In the end system, if the AAL layer receives a PDU whose sequence number does not directly follow that of the previous PDU received on that VC, the PDU just received shall be treated as if it had an invalid CRC. The the sequence number just received shall be remembered for comparison with the sequence number of the next PDU to be received on that VC.

In the CIF-AD, if the AAL layer receives a PDU whose sequence number does not directly follow that of the previous PDU received on that VC, the CRC field shall be filled with an intentionally invalid CRC and all cells in the PDU shall be forwarded. The sequence number just received shall be remembered for comparison with the sequence number of the next PDU to be received on that VC.

F.1.4 Other ATM Adaptation Layer Traffic

While AAL0, AAL1, and AAL5 are the only AALs guaranteed to work as required across CIF, there is nothing in CIF that would cause other new AALs not to work.

AAL1 is only expected to be used in the *simplified* form for voice interworking, and its use is discouraged. Instead, voice streams, which use 48-byte-filled AAL0 cells (over either VBR or ABR with MCR=64Kbps) and support voice activity detection, are preferred.

F.1.5 Available Bit Rate (ABR) Support

For all VCs using ABR traffic management, the CIF-AD shall act as the ABR source and destination, sending and receiving RM (resource management) cells on behalf of the end system. A simplified traffic management protocol shall be used between the CIF-AD and the end system, as described here.

The CIF-AD shall indicate to the end system the rate at which to send ABR traffic on a per-VC basis by sending the rate in in-rate RM cells. The rate shall be carried in the ER field of the RM cell. All other fields of the RM cell shall be ignored by the end system. This is the only type of RM cell (payload type = 6) supported across the Ethernet.

An RM cell may be generated by the CIF-AD to the end system whenever the ACR on an ABR VC changes significantly from the last time the ACR was sent, or periodically at a low rate. In the CIF-AD the ACR is normally updated when a backward RM cell is received from the network. It may also be updated when there is traffic in the forward direction but a backwards RM cell has not been received.

The exact policy as to when a CIF-AD sends RM cells to the end system, and how the rate at which the end system is permitted to send is derived, are not specified. Regardless of the rate at which the CIF-AD sends RM cells to the end system, it is recommended that the overall average ACR sent to the end system accurately reflect the average of the ACR actually received by the CIF-AD in RM cells from the network.

This specification does not require implementation of a mechanism by which the end system can control the flow of traffic from the CIF-AD (and the network beyond the CIF-AD) to it. An optional mechanism for doing so is as follows: the end system may manage the traffic flowing toward it on a particular VC by sending out-of-rate backward RM cells. If RM cells are sent, the CLP bit shall be 1, the direction shall be 1 (backward), the ER field shall be filled in, and the CI and/or NI bits may be set. If the control is being done with the CI or NI bits set to one, the ER field should be set to PCR. If control is being done with ER, then CI should be set to 0. The end system shall send no more than 10 RM cells per second. If the CIF-AD implements this option, it will need to save the ER, CI and NI state from these RM cells and use this state whenever it sends RM cells on this VC until such time as the end system clears the restriction by sending an RM cell with CI = 0, NI = 0, and ER greater than or equal to PCR. During this period, the CIF-AD shall use the ER from the end system as the maximum rate allowed on the VC and set CI or NI if these bits were set in the end system′s RM cell. The RM cell from the end system may be forwarded or not as desired.

When using ABR, the following applies:

- There is no mechanism by which the end system can control the flow of traffic flowing from the CIF-AD to the end system.
- The end system may manage the traffic flowing towards itself on a particular VC by sending out-of-rate backward RM cells. If RM cells are sent, the CLP bit shall be 1, the direction shall be 1 (backward), and the end system may send no more than 10 RM cells per second. The CIF-AD may process these RM cells, adjusting what it sends on the end system′s behalf based on their contents, or simply let them flow through into the network like any other cells.

F.1.6 Signalling

It is intended that ATM signalling functionality be fully supported without changes. The specific version(s) of signalling supported, at the end system or in the CIF-AD, is vendor-specific.

F.1.7 Management

F.1.7.1 ILMI Startup Procedures

Neither the CIF-AD nor the end system should need to have the other′s Ethernet MAC address configured. It is possible for both MAC addresses to be discovered dynamically. Both addresses must be known, as there may optionally be multiple devices on the Ethernet segment.

When the end system is starting up, it will have no knowledge of a CIF-AD′s MAC address. The end system sends ILMI poll messages, every *T* seconds, as described in ILMI. Poll messages are each sent in a single Ethernet frame. The MAC source address of the frame containing the poll is the sender′s, while the frame′s destination address is a MAC-layer multicast address allocated for CIF messages from end systems to CIF-ADs.

The MAC multicast address for the function described above has not yet been allocated.

If a CIF-AD has no ILMI connectivity established with any end system, and thus has no knowledge of any end system's MAC address, it does not send polls. Instead, it waits to hear a message, either multicast or unicast, from an end system. The CIF-AD responds to messages from all end systems.

When they have lost ILMI connectivity, end systems receive only unicast frames. CIF-ADs receive either multicast or unicast frames, but send only unicast frames to end systems from which they have heard. There is no scope for accidental establishment of ILMI connectivity between two CIF-ADs or two end systems.

Upon receipt of a response, the end system learns the MAC address of the CIF-AD. From this time on it will send ILMI messages, including polls, only in point-to-point unicast frames destined to the CIF-AD's MAC address. ColdStartTrap messages can be exchanged at this point.

When there are multiple CIF-ADs on the same Ethernet segment, an end system may receive responses from more than one. Before sending a coldStartTrap message, the end system shall choose a CIF-AD from among those that respond to its multicasts. After the choice is made, following the above rule, the end system shall send ILMI messages only to that CIF-AD. Messages from that CIF-AD shall be accepted, but messages from any other CIF-AD shall be ignored. Subsequently, if the end system enters the *connectivity lost* state again, it will again accept messages from any CIF-AD, and choose one to send its coldStartTrap messages to.

When there are multiple end systems on the same Ethernet segment, a CIF-AD may receive messages from more than one end system. If the CIF-AD supports multiple end systems on a segment (optional), it shall respond to each as it if had a point-to-point connection to it, that is with unicast point-to-point frames containing coldStartTrap messages.

As with any ATM interface, an end system may receive an ILMI message from the CIF-AD when it has lost connectivity but before it can send a (multicast) poll, or a CIF-AD may receive a unicast ILMI message that is not a poll. In that case it may assume connectivity is established, save the message's source MAC address and immediately send a coldStartTrap message in a unicast frame.

In this procedure, the CIF-AD's startup sequence is slightly modified from the usual ILMI sequence. Instead of sending unsolicited polls, the CIF-AD only sends polls upon the receipt of a poll from the end system.

The normal startup sequence for the end system is unchanged. Additional logic must be added to (1) send polls in multicast frames initially, and (2) to learn the Ethernet MAC address of the CIF-AD from the subsequent response sent by the CIF-AD.

Other ILMI Support: Except for the procedures defined in F.1.7.1, "ILMI Startup Procedures" on page 219, it is intended that all ATM ILMI functionality be fully supported without changes. The specific version(s) of ILMI supported, at the end system or in the CIF-AD, is vendor-specific.

The ILMI MIB shall be supplemented with the version number(s) of Cells-in-Frame supported across the link. All implementations shall support the initial version.

F.1.8 Discussion

The CIF document is changing rapidly and although some of the functions are well defined, there are many functions or problems that are being worked on. The following sections describe some of the open issues.

CIF-AD to CIF-ES Configurations

The expected configuration is a single end system sharing a half-duplex Ethernet segment with a single CIF-AD. Typically, this would be a 10BASE-T segment. Other Ethernet technology can be used (for example, 100BASE-T). It is possible to carry ATM over 802.3 and 802.5 networks, as well as using the same mechanisms, but the CIF document only concerns itself with Ethernet Version 2.

100 Mbps Ethernet

On half-duplex Ethernet segments, in order to minimize jitter, it is recommended that the CIF-AD implement some mechanism to avoid the Ethernet *capture effect* in which one sender transmits all of its queued frames before the other can gain control of the Ethernet and transmit any.

No timing reference is carried across the Ethernet. If synchronization is required, a mechanism should be used at a layer above CIF, which does not depend on a physical layer timing reference.

Multiple CIF-ADs may be supported on a single segment, for redundancy. The end system shall discover and select a CIF-AD as specified in F.1.7.1, "ILMI Startup Procedures" on page 219. The procedure for selecting a CIF-AD is strictly an end system issue, although it is expected that end systems shall simply pick the first CIF-AD which responds.

Multiple end systems may share a segment, but doing so is discouraged. In such a configuration there can be no quality-of-service guarantees. However, since UBR and ABR (MCR=0) require no QOS, they may be used to provide services such as Classical IP and LAN Emulation to multiple end systems on a single segment.

Except for startup ILMI messages, all traffic from a CIF-AD to an end system is normally point-to-point unicast. However, in the case of multiple end systems per Ethernet segment, point-to-multipoint traffic may optionally be sent out as

MAC-level multicast frames. In this configuration, end systems shall discard multicast frames not sent from their active CIF-AD, as well as multicast frames sent to inactive VCs.

Point-to-multipoint traffic might be replicated on segments with multiple end systems if it originates at one end system and another is a recipient. Therefore, high volume and/or delay sensitive point-to-multipoint traffic requires the expected configuration of a single end system per segment. A configuration with multiple end systems on a single segment would allow low-volume, delay-insensitive point-to-multipoint traffic.

The CIF Attachment Device (CIF-AD)

The CIF-AD may be an ATM switch, or it may be a simple ATM multiplexor or concentrator, as long as it performs the CIF mechanisms correctly. That is, in order to reduce complexity in the CIF-AD, traffic between two end systems attached to the CIF-AD is allowed to have to travel through a CIF-AD concentrator to an ATM switch and back. Such possibilities are outside the scope of this document.

On behalf of end systems that are not yet CIF-capable, the CIF-AD may run LAN Emulation and/or MPOA clients. CIF-capable end systems will run their own such clients.

For ease of transition from existing Ethernet-based protocols to ATM, the CIF-AD may support two modes of Ethernet traffic on an Ethernet: CIF and all other traffic. The CIF-AD may switch from legacy traffic to CIF automatically when it sees frames with the CIF Ethernet type. In that way the users can switch from legacy Ethernet-based operation to CIF, on a per-port basis, when they are ready. Ports transmitting legacy traffic may be treated as a common repeated or bridged Ethernet or switched.

The CIF-AD may, but need not, support mixing of CIF frames and legacy Ethernet frames on the same segment. The CIF-AD may require that a port either use CIF or legacy Ethernet exclusively. If CIF frames and legacy Ethernet frames are allowed to be mixed on the same segment, care should be taken not to interfere with ATM traffic for which QOS agreements have been made. If no other queueing policies are in use outside of ATM, the legacy frames should be given no higher priority than ATM UBR traffic.

The CIF End System (ES)

Within an end system the existing Ethernet driver and higher layer software can remain intact. *Shims* built using standard techniques can be placed just above the Ethernet driver, or the Ethernet driver can be replaced.

The AAL layer will need slight enhancement (for example, in AAL5 CRC handling). Just as in any other ATM implementation, legacy applications will use a system- and protocol-specific interface, such as LAN Emulation, to access the ATM services.

This very same picture could be used to show the internal organization of an end system with an ATM-specific interface card. Above the AAL SAP, a CIF interface will be indistinguishable from any other type of ATM interface.

Frames

In order to support quality of service, both the end system and the CIF-AD should implement multiple queues and weighted fair queuing. In the end system, queue management can be in the CIF shim, just above the Ethernet driver. Preferred methods of queue management and traffic shaping are not a CIF protocol issue and are only given minimal attention here. In general, low-bandwidth but delay-sensitive traffic such as AAL1 voice should be carried in frames containing only one cell, while high-bandwidth but delay-tolerant traffic such as most AAL5 streams should have maximally filled frames (to minimize the end system's interrupt rate). It should be possible for either the CIF-AD or the end system to send a frame containing delay-sensitive data even while building a frame to send delay-tolerant data. Delay-sensitive data should not be held up while delay-tolerant data is accumulated.

If the CIF-AD and the end system allow transmission of non-CIF frames along with CIF frames, the non-CIF frames should pass through queue management and be given no better treatment than ATM UBR traffic, to avoid interference with ATM QOS.

There is no means to negotiate the number of SAR-PDUs per frame.

Error detection

The HEC field is not needed to protect cell payloads across the Ethernet, and thus does not appear in the CIF header, since each Ethernet frame is protected by the Ethernet frame CRC.

The only weakness introduced by not using the AAL5 CRC across the Ethernet, depending instead on the Ethernet CRC, is that in poorly installed Ethernets there is a chance of *packet splices*, where an apparent single PDU is received which is really a combination of two actual PDUs. Without any additional protection, the probability of an undetected packet splice being delivered to the AAL5 client is O((p•)/2), where *p* is the probability of an Ethernet frame being lost and not retransmitted. However, use of the 4-bit AAL5 PDU sequence number as described in F.1.3.3, "The PDU Sequence Number" on page 217 completely detects packet splices. The possibility of errors going undetected by the Ethernet CRC still remains, with the same probability as today, about O(1e-10).

Appendix G. Server Cache Synchronization Protocol (SCSP) -NBMA

25 on page 231 describes the Server Cache Synchronization Protocol (SCSP) for NBMA networks. The document is an Internet draft and is changing continually.

SCSP attempts to solve the generalized server synchronization/cache- replication problem wherein a set of server entities that are bound to a server group (SG) through some means (for example, all servers belonging to the same Logical IP Subnet (LIS)) wish to synchronize the contents (or a portion thereof) of their caches. These caches contain information on the state of the clients within the scope of interest of the SG. An example of types of information that must be synchronized can be seen in NHRP using IP where the information includes the registered client's IP to NBMA mappings in the SG LIS.

If accepted as a standard, SCSP will become the mechanism of choice used by various protocols (for example, NHRP, ATMARP, MARS, etc.) to synchronize caches across multiple platforms.

Only the first few pages of the draft document constitute the SCSP description proper. However, the document also includes a description of the use of SCSP by a number of protocols (for example, NHRP, ATMARP, MARS, etc.) and some optional functionality that may be implemented as deemed appropriate. The authors hope that the appendices to the document will spark interest in applying SCSP to the server synchronization needs of other protocols by supplying examples of SCSP's use.

Appendix H. Special Notices

This publication is intended for system engineers, networking consultants, network designers, and network administrators who need to understand IBM's switched virtual networking architecture. The information in this publication is not intended as the specification of any programming interfaces that are provided by SVN products. See the PUBLICATIONS section of the IBM Programming Announcement for SVN products for more information about what publications are considered to be product documentation.

References in this publication to IBM products, programs or services do not imply that IBM intends to make these available in all countries in which IBM operates. Any reference to an IBM product, program, or service is not intended to state or imply that only IBM's product, program, or service may be used. Any functionally equivalent program that does not infringe any of IBM's intellectual property rights may be used instead of the IBM product, program or service.

Information in this book was developed in conjunction with use of the equipment specified, and is limited in application to those specific hardware and software products and levels.

IBM may have patents or pending patent applications covering subject matter in this document. The furnishing of this document does not give you any license to these patents. You can send license inquiries, in writing, to the IBM Director of Licensing, IBM Corporation, 500 Columbus Avenue, Thornwood, NY 10594 USA.

The information contained in this document has not been submitted to any formal IBM test and is distributed AS IS. The information about non-IBM ("vendor") products in this manual has been supplied by the vendor and IBM assumes no responsibility for its accuracy or completeness. The use of this information or the implementation of any of these techniques is a customer responsibility and depends on the customer's ability to evaluate and integrate them into the customer's operational environment. While each item may have been reviewed by IBM for accuracy in a specific situation, there is no guarantee that the same or similar results will be obtained elsewhere. Customers attempting to adapt these techniques to their own environments do so at their own risk.

The following terms are trademarks of the International Business Machines Corporation in the United States and/or other countries:

Advanced Peer-to-Peer Networking	APPN
IAA	IBM
Nways	RT

The following terms are trademarks of other companies:

C-bus is a trademark of Corollary, Inc.

PC Direct is a trademark of Ziff Communications Company and is used by IBM Corporation under license.

UNIX is a registered trademark in the United States and other countries licensed exclusively through X/Open Company Limited.

Microsoft, Windows, and the Windows 95 logo are trademarks or registered trademarks of Microsoft Corporation.

AppleTalk	Apple Computer, Incorporated
Bell	AT&T Bell Laboratories Incorporated
Crosstalk	Digital Communications Associates, Incorporated
DECnet	Digital Equipment Corporation
Digital	Digital Equipment Corporation
IPX	Novell, Incorporated
NDIS	3Com Corporation and Microsoft Corporation
Novell	Novell, Incorporated
System 7	Apple Computer, Incorporated
Xerox	Xerox Corporation

Other trademarks are trademarks of their respective companies.

Appendix I. Bibliography

The publications listed in this section are considered particularly suitable for a more detailed discussion of the topics covered in this book.

1. The Rate-Based Flow Control Framework for the ABR ATM Service

   Bonomi, F., Fendick, K.W.
   IEEE Network
   March/April 1995

2. Address Resolution Protocol

   Plummer, D.C.
   RFC 826
   1982

3. Classical IP and ARP over ATM

   Laubach, M.
   RFC 1577
   1994

4. Integrated Services in the Internet Architecture: An Overview

   Braden, R., Clark, D., Shenker, S.
   RFC 1633
   1994

5. IP over ATM: A Framework Document

   Cole, R., Shur, D.,Villamizar, C.
   RFC 1932
   1996

6. Transmission of IP Datagrams over the SMDS Service

   Lawrence, J., Piscitello, D.
   RFC 1209
   1991

7. Multiprotocol Encapsulation over ATM Adaptation Layer 5

   Heinanen, J.
   RFC 1483
   1994

8. NBMA Next Hop Resolution Protocol (NHRP)

   Katz, D., Piscetello, D., Cole, B., Luciano, J.V.
   draft-ietf-rolc-nhrp-07.txt.

1995

9. ATM User-Network Interface (UNI) Specification Version 3.1

   ATM Forum

10. ATM Forum - Baseline Text for MPOA

   Draft ATM Forum/95-0824r6
   1995

11. Support for Multicast over UNI 3.0/3.1 based ATM Networks

   Armitage, G.
   <draft-ietf-ipatm-ipmc-12.txt>
   1996

12. NBMA Address Resolution Protocol (NARP)

   Heinanen, J.
   RFC 1735
   1994

13. Resource Reservation Protocol (RSVP), Version 1 Functional Specification

   Braden, R., Zhang, L., Berson, S., Herzog, S.
   Jamin, S.
   <draft-ietf-rsvp-spec-10.txt>
   1996

14. Internet Protocol, Version 6 (IPv6), Specification

   Deering, S., Hinden, R.
   RFC 1883
   1996

15. A Framework for IPv6 over ATM

   Schulter, P.
   <draft-schulter-ipv6atm-framework-01.txt>
   1996

16. IPv6 Neighbor Discovery over ATM

   Armitage, A.
   <draft-ietf-ipatm-ipv6nd-02>
   1996

17. LAN Emulation over ATM Version 2

   Draft ATM Forum/95-1082R2
   1996

18. APPN Architecture and Product Implementations Tutorial

Lenhard, P., Purrington, J., Pickering, R.
GG24-3669-02
April 1994

19. Networking BroadBand Services (NBBS) Architecture Tutorial

Lenhard, P., Kelly, J., Pickering, R.
GG24-4486-00
June 1995

20. APPN Architecture Version 2

IBM
SC30-3422-03

21. High-speed Networking Technology: An Introductory Survey

Lenhard, P., Dutton, H.J.R.
Prentice Hall, ISBN 0-13-242421-5
1995

22. ATM Forum LAN Emulation over ATM Version 1.0 Addendum

1996

23. Private Network-Network Interfaces Specification Version 1.00

Draft ATM Forum/94-0471R7
1996

24. Cells In Frames Version 1.0: Specification, Analysis and Discussion

Brim, S.W., Cogger, R., Hill, G., Kumar, S.,
Roberts, L., Yang, J.

<http://cif.cornell.edu/specs/V1.0/CIF-baseline.html>
1996

25. Server Synchronization Protocol (SCSP) - NBMA

Armitage, G., Halpern, J., Luciani, J.V.
<draft-luciani-rolc-scsp-02.txt>
1996

List of Abbreviations

A

AAL ATM Adaptation Layer

ABR Available Bit Rate

ACM Address Complete Message

ACT Activity Bit

ADPCM Adaptive Differential Pulse Code Modulation

AFI Authority and Format Identifier: first byte of the ATM address that determines the address type. AFI39 is DCC, 47 is ICD, and 45 indicates an E.164 format.

AHFG ATM-attached Host Functional Group (MPOA SWG)

Ai Signalling ID assigned by Exchange A

AII Active Input Interface (Used in UNI PMD specs for Copper/Fiber)

AIM ATM Inverse Multiplexer

AIR Additive Increase Rate

AIS Alarm Indication Signal (UNI Fault Managment)

AIS-E Alarm Indication Signal - External

AMI Alternate Mark Inversion

ANI Automatic Number Identification

ANM Answer Message

ANSI American National Standards Institute

AOI Active Output Interface (Used in UNI PMD specs for Copper/FIber)

API Application Programming Interface

APPN Advanced Peer to Peer Network

ARE All Routes Explorer

ARP Address Resolution Protocol

ARQ Automated Repeat reQuest

ASE Application Service Element

ASIC Application Specific Integrated Circuit

ASN Abstract Syntax Notation

ASN.1	Abstract Syntax Notation One
ASP	Abstract Service Primitive
ATD	Asynchronous Time Division
ATE	ATM Terminating Equipment (SONET)
ATM	Asynchronous Transfer Mode
ATS	Abstract Test Suite
ATM	Asynchronous Transfer Mode
ATMARP	ATM Address Resolution Protocol
AUU	ATM User-to-User
AVSSCS	Audio-Visual Service Specific Convergence Suplayer (ATM Forum)

B

B-HLI	Broadband High Layer Information
B-ICI	Broadband Inter Carrier Interface
B-ICI SAAL	B-ICI signalling ATM Adaptation Layer
B-ISDN	Broadband Integrated Services Digital Network
B-ISUP	Broadband ISDN User's Part
B-ICI	Broadband Intercarrier Interface
B-ISSI	Broadband Inter-Switching System Interface
B-LLI	Broadband Low Layer Information
B-LLI	Broadband Low Layer Information
B-NT	Broadband Network Termination
B-TE	Broadband Terminal Equipment
BBC	Braodband Bearer Capability
BCBDS	Broadband Connectionless Data Bearer Service
BCD	Binary Coded Decimal
BCOB	Broadband Class of Bearer
BECN	Backward Explicit Congestion Notification
BER	Basic Encoding Rules (ASN-1)
BER	Bit Error Rate (link quality specification/testing)
BGP	Border Gateway Protocol
BGT	Broadcast and Group Translators

Bi	Signalling ID assigned by Exchange B
BIP	Bit Interleaved Parity (e.g. SONET BIP-8 for paht error monitoring)
BIPV	Bit Interleaved Parity Violation
BIS	Border Intermediate System (ATM Forum, PNNI SWG)
BISDN	Broadband - Integrated Services Digital Network
BISSI	Broadband Inter Switching System Interface
BN	Bridge Number
BOF	Birds of Feather
BOM	Begining of Message
BOOTP	Bootstrap Protocol
BPDU	Bridge Protocol Data Unit
BPP	Bridge Port Pair
BPS	Bits per second
BSS	Broadband Switching System
BSVC	Broadcast Switched Virtual Connections
BT	Burst Tolerance
BT	Begin Tag
BUS	Broadcast and Unknown Server
BW	Bandwidth

C

CA	Cell Arrival
CAC	Connection Admission Control
CACCAC	Connection Admission Control
CBDS	Connectionless Broadband Data Service
CBR	Constant Bit Rate
CBR interactive	Constant Bit Rate interactive
CBR noninteractive	Constant Bit Rate non interactive
CC	Continuity Cell
CCITT	Consultative Committee on International Telephone & Telegraph
CCR	Current Cell Rate
CCS	Common Channel Signalling
CCSS7	Common Channel Signalling System 7
CDT	Cell Delay Tolerance
CDV	Cell Delay Variation
CDVT	Cell Delay Variation Tolerance

CEI	Connection Endpoint Identifier
CER	Cell Error Ratio
CES	Circuit Emulation Service
CI	Congestion Indicator
CIDR	Classless Inter-Domain Routing
CIP	Carrier Identification Parameter
CIR	Committed Information Rate
CL	Connectionless
CLNAP	Connectionless Network Access Protocol
CLNP	Connectionless Network Protocol
CLNS	Connectionless Network Service
CLP	Cell Loss Priority
CLR	Cell Loss Ratio
CLS	Connectionless Server
CLSF	Connectionless Service Function
CME	Component Management Entity
CMI	Coded Mark Inversion
CMISE	Common Management Information Service Element
CMIP	Common Management Interface Protocol
CMR	Cell Misinsertion Rate
CN	Copy Network
CNM	Customer Network Management
CNR	Complex Node Representation (ATM Forum, PNNI SWG)
CO	Connection Oriented
COD	Connection Oriented Data
COM	Continuation of Message
COS	Class of Service
CP	Connection Processor
CPCS	Common Part Convergence Sublayer
CPE	Customer Premise Equipment
CPG	Call Progress Message
CPN	Customer Premises Network
CPN	Calling Party Number
CPI	Common Part Indicator
CRC	Cyclic Redundance Check

CRCG	Common Routing Connection Group
CRF(VC)	Virtual Channel Connection Related Function (related ot UPC/UNI 3.0)
CRF(VP)	Virtual Path Connection Related Function (related ot UPC/UNI 3.0)
Crankback IE	Crankback - Information Element
CRS	Cell Relay Service
CS	Convergence Supblaye (as in CS_PDU)
CS	Carrier Selection
CS1	Capability Set One
CS2	Capability Set Two
CSI	Convergence Sublayer Indication
CSPDN	Circuit Switched Public Data Network
CSR	Cell Missequenced Ratio
CSU	Channel Service Unit
CTD	Cell Transfer Delay
CTV	Cell Tolerance Variation

D

DA	Destination MAC address
DA	Destination Address
DCC	Data Country Code
DCE	Data Communication Equipment
DD	Depacketization Delay
DLC	Data Link Control
DES	Destination End System
DLCI	Data Link Connection Identifier
DMDD	Distributed Multiplexing Distributed Demultiplexing
DN	Distribution Network
DQDB	Distributed Queue Dual Bus
DS	Distributed Single Layer Test Method
DS-0	Digital Signal, Level O
DS-1	Digital Signal, Level 1
DS-2	Digital Signal, Level 2
DS-3	Digital Signal, Level 3

DS3 PLCP	Physical Layer Convergence Protocol
DSE	Distributed Single Layer Embedded Test Method
DSID	Destination Signalling Identifier
DSS2	Setup Digital Subscriber Signalling #2
DSU	Data Service Unit
DSX	Digital Signal Cross-Connect
DTE	Data Terminal Equipment
DTL IE	DTL - Information Element
DXI	Data Exchange Interface

E

EDFG	Edge Device Functional Group (ATM Forum, MPOA SWG)
EFCI	Explicit Forward Congestion Indication
ELAN	Emulated LAN (ATM Forum LANE)
EMI	Electromagnetic Interference
EOM	End of Message
EPRCA	Enhanced Proportional Rate Control Algorithm (ATM Forum)
ETAG	ESI
End Station Identifier	End Tag
ETE	End-to-End
EXM	Exit Message

F

FC	Feedback Control
FCS	Fast Circuit Switching
FCS	Frame Check Sequence
FCVC	Flow Controlled Virtual Circuit
FDDI	Fiber Distributed Data Interface
FEA	Functional Entity Action (UNI 3.0, C.3.2.3)
FEBE	Far End Block Error (SONET)
FEC	Forward Error Correction
FECN	Forward Explicit Congestion Notification
FERF	Far End Receive Failure
FG	Functional Group (ATM Forum, MPOA SWG)
FUNI	Frame Based Use-to-Network Interface (ATM Forum)

FRS	Frame Relay Service
FUNI	Frame User Network Interface

G

GAP	Generic Address Parameter
GCID	Global Call Identifier
GCID-IE	Global Call Identifier-Information Element
GCRA	Generic Cell Rate Algorithm
GFC	Generic Flow Control
GRC	Generic Reference Configuration

H

HBFG	Host Behavior Functional Group (ATM Forum, MPOA SWG)
HDB3	High Density Bipolar 3
HDLC	High Level Data Link Control
HEC	Header Error Control
HEC	Header Error Check
HEL	Header Extension Length
HLPI	Higher Layer Protocol Identifier
HOL	Head of Line

I

IAA	Initial Address Acknowledgment
IAM	Initial Address Message
IAR	Initial Address Reject
IASG	Internetwork Address Sub-Group (ATM Forum, MPOA SWG)
IBSG	Internetwork Broadcast Sub-Group (ATM Forum, MPOA SWG)
IBUFG	Internetwork Broadcast/Unknown Functional-Group (ATM Forum, MPOA SWG)
IC	Initial Cell Rate
ICD	International Code Designator
ICFG	IASG Coordination Function Group (ATM Forum, MPOA SWG)
ICMP	Internet Control Message Protocol
IDU	Interface Data Unit
IE	Information Element
IEC	Interexchange Carrier

IEEE Institute of Electrical and Electronics Engineers

IETF Internet Engineering Task Force

IISP Interim Inter-Switch Protocol akd. P-NNI Phase 0

ILMI Interim Link Management Interface

ILMI Interim Local Management Interface

IOP Interoperability

IP Internet Protocol

Ipng Internet Protocol Next Generation

IPX Novell Internetwork Packet Exchange

ISO International Organization for Standardization

ITU International Telecommunications Union

IUT Implementation Under Test

IWF Interworking Function

IWU Interworking Unit

J

JPEG Joint Photographic Experts Group

L

LAN Local Area Network

LANE Local Area Network Emulation (ATM Forum)

LAPD Link Access Procedure D

LB Leaky Bucket

LCD Loss of Cell Delineation

LCT Last Compliance Time (Used in GCRA definition)

LD LAN Destination

LE LAN Emulation (also, LANE)

LE_ARP LAN Emulation Address Resolution Protocol

LEC LAN Emulation Client

LEC Local Exchange Carrier

LECID LAN Emulation Client Identifier

LECS LAN Emulation Configuration Server

LES LAN Emulation Server

LGN Logical Group Node (ATM Forum, PNNI)

LIJP Leaf Initiated Join Parameter

LIS	Logical IP Subnet (rfc 1577)
LIV	Link Integrity Verification
LLATMI	Lower Layer ATM Interface
LLC	Logical Link Control
LLC/SNAP	Logical Link Control/Subnetwork Access Protocol
LMI	Layer Management Interface
LOC	Loss of Cell delineation
LOF	Loss of Frame (UNI Fault Management)
LOP	Loss of Pointer (UNI Fault Management)
LOS	Loss of Signal (UNI Fault Management)
LSB	Least Significant Bit
LSR	Leaf Setup Request
LTE	Line Terminating Equipment (SONET)
LTLT	Lower Tester
LTHLTH	Length Field
LUNI	LANE UNI (ATM Forum, see LANE)

M

MAMA	Maintenance and Adaptation
MAC	Medium Access Control
MAN	Metropolitan Area Network
MARS	Multicast Address Resolution Service (Draft IETF - IPATM)
MBS	Maximum Burst Size
MCR	Minimum Cell Rate
MCTD	Mean Cell Transfer Delay
ME	Mapping Entity
MIB	Management Information Base
MID	Message Identifier
MIN	Multistage Interconnection Networks
MIR	Maximum Information Rate
MMF	Multimode Fiberoptic cable
MPEG	Motion Picture Experts Group
MPOA	Multiprotocol over ATM (ATM Forum)
MRCS	Multirate Circuit Switching
MS	Meta Signalling
MSAP	Management Service Access Point

MSB	Most Significant Bit
MSN	Monitoring Cell Sequence Number
MSVC	Meta-signalling Virtual Channel
MT	Message Type
MTP	Message Transfer Part
MTU	Message Transfer Unit

N

N-ISDN	Narrowband Integrated Services Digital Network
NBMA	Nonbroadcast Multiple Access
NDIS	Network Driver Interface Specification
NE	Network Element
NEBIOS	Network Basic Input/Output System
NEXT	Near End Crosstalk
NHRP	Next Hop Routing Protocol (from IETF ROLC WG)
NHS	Next Hop Server
NIU	Network Interface Unit
NLPID	Network Layer Protocol Identifier
NMS	Network Management System
NNI	Network to Network Interface
NP	Network Performance
NPC	Network Parameter Control
NRM	Network Resource Management
NSAP	Network Service Access Point
NSAPA	Network Service Access Point Address
NSP	Network Service Provider
NSR	Nonsource Routed
NT	Network Termination

O

OAM	Operations, Administration and Maintenance
OCD	Out-of-Cell Delineation (UNI 3.0 section 2.1.2.2.2)
ODI	Open Data-Link Interface
OLI	Originating Line Information
OOF	Out of Frame
OPCR	Original Program Clock Reference
OSI	Open systems Interconnection

OSID	Origination Signalling Identifier
OSPF	Open Shortest Path First
OUI	Organization Unique Identifier
OUI	Organizational Unit Identifier

P

P-NNI	Private Network to Network Interface
PAD	Packet Assembler and Disassembler
PBX	Private Branch eXchange
PC	Priority Control
PC	Protocol Control
PCM	Pulse Code Modulation
PCO	Point of Control and Observation
PCR	Peak Cell Rate (UNI 3.0)
PCR	Program Clock Reference
PCVS	Point to Point Switched Virtual Connections
PD	Packetization Delay
PDH	Plesiochronous Digital Hierarchy
PDU	Packet Data Unit
PDU	Protocol Data Unit
PGL	Peer Group Leader (ATM Forum, PNNI)
PHY	Physical Layer of the OSI model
PHY	Physical Layer
PICS	Protocol Implementation Conformance Statement
PID	Protocol Identifier Governing Connection Types
PIXIT	Protocol Implementation eXtra Information for Testing
PL-OU	Physical Layer Overhead Unit (UNI physical layer frame definition)
PL	Physical Layer
PLL	Phase Locked Loop
PLCP	Physical Layer Convergence Protocol
PM	Physical Medium
PMD	Physical Layer Dependent sublayer
PMP	Point to Mulitpoint (UNI 3.0)
PNNI	Private Network Node Interface (ATM Forum, PNNI SWG)

PNNI	Private Network-to-Network Interface (ATM Forum, PNNI SWG)
POH	Path Overhead
POI	Path Overhead Indicator
PT	Payload Type
PTE	Path Terminating Equipment (SONET)
PTI	Payload Type Identifier
PTSE	PNNI Topology State Element (ATM Forum, PNNI)
PTSP	PNNI Topology State Packet (ATM Forum, PNNI)
PVC	Permanent Virtual Circuit
PVCC	Permanent Virtual Channel Connection
PVPC	Permanent Virtual Path Connection

Q

QD	Queuing Delay
QOS	Quality of Service
QPSX	Queue Packet and Synchronous Circuit Exchange

R

RAI	Remote Alarm Indication
RBOC	Regional Bell Operating Company
RC	Routing Control
RD	Route Descriptor
RDF	Rate Decrease Factor
RDI	Remote Defect Identification (UNI Fault Management)
RDI	Remote Defect Indication
REL	Release Message
RFC	Request For Comment (Document Series)
RFI	Radio Frequency Interference
RI	Routing Information
RII	Routing Information Indicator
RIP	Routing Information Protocol
RISC	Reduced Instruction Set Computing
RLC	Release Complete
RM	Resource Management
ROLC	Routing Over Large Clouds
RSFG	Route Server Functional Group (ATM Forum, MPOA SWG)

RSVP (protocol)	Resource Reservation Protocol
RT	Routing Type
RTS	Residual Time Stamp

S

SA	Source MAC address
SA	Source Address
SAAL	Signalling ATM Adaptation Layer
SAP	Service Access Point
SAR	Segmentation and Reassembly
SCCP	Signalling Connection and Control Part
SCP	Service Control Point
SCR	Sustainable Cell Rate (UNI 3.0)
SDH	Synchronous Digital Hierarchy
SDU	Service Data Unit (as in AAL SDU)
SE	Switching Element
SEAL	Simple and Efficient Adaptation Layer
SECB	Severely Errored Cell Block
SF	Switching Fabric
SGM	Segmentation Message
SID	Signalling Identifier
SIPP	SMDS Interface Protocol
SIR	Sustained Information Rate
SMC	Sleep Mode Connection
SMDS	Switched Multimegabit Data Services
SMF	Single Mode Fiber
SN	Sequence Number
SNA	Systems Network Architecture
SNAP	Sub Network Access Protocol
SNDCF	Subnetwork Dependent Convergence Function (ATM Forum, MPOA SWG)
SNI	Subscriber Network Interface
SNMP	Simple Network Management Protocol
SOH	Section Overhead
SONET	Synchronous Optical Network
SPID	Service Protocol Identifier
SPTS	Single Program Transport Stream
SR	Source Routing (Bridging)
SRF	Specifically Routed Frame

SRT	Source Routing Transparent
SRTS	Synchronous Residual Time Stamp
SSCF	Service Specific Coordination Function
SSCOP	Service Specific Connection Oriented Protocol
SSCS	Service Specific Convergence Sublayer
ST	Segment Type
STE	Spanning Tree Explorer
STM	Synchronous Transfer Mode
STM1	Synchronous Transport Mode 1 -- 155mbits/sec
STP	Signalling Transfer Point
STP	Shielded Twisted Pair cable
STS	Synchronous Time Stamps
STS-3c	Synchronous Transport System-Level 3 concatenated
SUT	System Under Test
SVC	Switched Virtual Circuit
SVCI	Switched Virtual Circuit Identifier
SVP	Switched Virtual Path
SWG	Subworking Group

T

T1S1	ANSI T1 Subcommittee
TAT	Theoretical Arrival Time (used in GCRA definition)
TAXI	Transparent asynchronous transmitter/receiver interface
TB	Transparent Bridging
TC	Transaction Capabilities
TC	Transmission Convergence
TCAP	Transaction Capabilities Applications Part
TCI	Test Cell Input
TCO	Test Cell Output
TCP	Transmission Control Protocol
TCP	Test Coordination Procedure
TCP/IP	Transmission Control Program/Internet Protocol
TCS	Transmission Convergence Sublayer
TDJ	Transfer Delay Jitter
TDM	Time Division Multiplexing

TE	Terminal Equipment
TIG	Topology Information Group (ATM Forum, PNNI)
TLV	Type / Length / Value
TM	Traffic Management
TM SWG	Traffic Management Subworking Group
TMP	Test Management Protocol
TNS	Transit Network Selection
TPCC	Third Party Call Control
TS	Traffic Shaping
TS	Time Stamp
TS	Transport Stream
TS	Time Slot
TSAP	Transport Service Access Point
TUC	Total User Cell count
TUCD	Total User Cell Difference

U

UBR	Unspecified Bit Rate
UDP	User Datagram Protocol
UME	UNI Management Entity (used in ILMI definition)
UNI	User Network Interface
UPC	Usage Parameter Control
UT	Upper Tester
UTOPIA	Universal Test & Operations PHY Interface for ATM
UTP	Unshielded Twisted Pair cable

V

VBR	Variable Bit Rate
VBR delay sensitive	Variable Bit Rate delay sensitive
VBR delay tolerant	Variable Bit Rate delay tolerant
VBR noninteractive	Variable Bit Rate noninteractive
VC	Virtual Channel (Virtual Circuit)
VC-Multiplexing	Virtual Channel - Multiplexing
VCC	Virtual Channel Connections
VCI	Virtual Connection Identifier
VCI	Virtual Channel Identifier
VCL	Virtual Channel Link (UNI 3.0)
VINCE	Vendor Independent Network Control Entity
VLAN	Virtual Local Area Network

VP Virtual Path

VP/VC Virtual Path, Virtual Circuit

VPC Virtual Path Connection

VPCI/VCI Virtual Path Connection Identifier/Virtual Channel Identifier

VPI Virtual Path Identifier

VPL Virtual Path Link (UNI 3.)

VPT Virtual Path Terminator (UNI 3.)

VS Virtual Scheduling

W

WAN Wide Area Network

WAN Wide Area Network

X

XNS Xerox Network Systems

XTP eXpress Transport Protocol

Index

Numerics

A

B

C

D

E

M

N

O

P

Q

R

S

T

U

V

W

SG24-4699-00